Tuning the Mind

Geometries of Consciousness

Holonomic Brain Theory
And
The Implicate Order

Shelli Renée Joye

B.S. Electrical Engineering
M.A. Indian Philosophy
Ph.D. Consciousness Studies

The Viola Institute Inc.
Viola, California

Dedicated to David Bohm and Karl Pribram,
and to all my teachers and mentors, including
Haridas Chaudhuri,
Ramamurti S. Mishra,
Allan Watts
and to my dissertation committee:
Brian Swimme, Allan Combs, Dean Radin;
and especially to my dear ones Alyssa, Jason, and Teresa,
and most especially to Susanne Rohner
who with loving kindness helped me produce this book.

Published in the United States of America
by The Viola Institute Inc.
35366 Wild Turkey Lane
Shingletown, California, USA
www.violainstitute.com

Printed in the United States of America

Tuning the Mind:
Geometries of Consciousness. -- 1st ed.
ISBN-13: 978-0-9988785-1-5 (Pbk)
ISBN-10: 0998878510

"Shelli Joye's integrative masterpiece, *Tuning the Mind*, reads like insane science fiction—quantum black holes, a universal center that's everywhere and nowhere, the dark energy holoflux, and of course, the mysterious Teilhardian isospheres. But the strange thing is, it's not insane, nor is it fiction. It's a novel and refreshing scientific take on the role of consciousness in the physical world. Well worth the read."

—**DEAN RADIN, Ph.D.**,
Chief Scientist, Institute of Noetic Sciences (IONS), and author of
Supernormal: Science, Yoga, and the
Evidence for Extraordinary Psychic Abilities

"Dr. Joye has not only created a map of consciousness; she has created a map of reality. Everything we call reality is experienced in consciousness, known in consciousness, and is in fact a modified form of consciousness. Reading this book will move you beyond physical materialism into better understanding the realities of consciousness."

—**DEEPAK CHOPRA, M.D.**, endocrinologist and author of
You are the Universe with the physicist Menos Kafatos

"Not since John Lilly has writing about consciousness exhibited such insight, clarity, and scientific acumen. Dr. Joye has given us a work without equal. It is a truly exciting and deeply informative read."

—**ALLAN COMBS, Ph.D.**
Director, Consciousness Studies Center, CIIS, author of
Consciousness Explained Better and *The Radiance of Being*

"Shelli Joye's book joins a growing stream of studies that focus on the new understanding of consciousness as a real and significant part of the world, using advanced concepts from quantum physics, holography and the related energy fluxes. It makes a signal contribution to the advance of the field I have called new paradigm consciousness research, and merits the attention of all open and forward looking investigators of the deep dimension of the world, of which consciousness is now recognized to be a principal element."

—**ERVIN LÁSZLÓ, Ph.D.**, systems theorist and author of
The Akashic Experience and *Science and the Akashic Field*

"Insightful, articulate, authoritative, and thought-provoking in the approach to understanding the physics of consciousness. I highly recommend reading *Tuning the Mind: Geometries of Consciousness* to anyone interested in the physics of consciousness. The clearly written discussions of the work of Karl Pribram and David Bohm are invaluable and set the groundwork for a deeper understanding of heart coherence and heart intelligence."

—**ROLLIN MCCRATY, Ph.D.**
Director of Research, HeartMath Institute, author of
The Coherent Heart: Heart-Brain Interactions

"Over the past four centuries, physics, math, evolutionary biology, genetics, and computer science have all had their periods of intense research. But now in the twenty-first century, it is beginning to look as if brain science and consciousness studies will become the new center of scientific research. As with the exploration of any new domain, hundreds of different theories are being advanced which tend to fall into half a dozen qualitatively different approaches. One of the most potent of these is the Pribram-Bohm lineage. Of the various works within this lineage, none is more exciting nor as comprehensive as Shelli Joye's: *Tuning the Mind: Geometries of Consciousness*. Her work is destined to become a classic in the field and is highly recommended to anyone searching for a way into this the cutting edge research."

—**BRIAN THOMAS SWIMME, Ph.D.**, Professor,
Philosophy, Cosmology, and Consciousness Studies,
California Institute of Integral Studies,
author of *The Universe Story* with Thomas Berry

"Based on her personal experience of a variety of conscious states throughout her exploratory life, and significant scientific research in neurophysiology, physics, engineering, and Eastern writings, Shelli has built a bridge connecting them all into a modern functional description of consciousness. Her ideas are a significant addition to the literature, and will encourage provocative thought and deep reflection."

—**JOHN TAYLOR, Ph.D.**
Dean, Orange Coast College

Introduction

This book is about consciousness and how it is structured, and how it may be navigated. Presented here is a psychophysical map, a guide for 21st century psychonauts (those highly inquisitive individuals motivated to explore altered states of awareness, states attained through natural ability, or through drugs, prayer, meditation, dreams, religious rituals, or physical exercises). I have been exploring non-ordinary noetic dimensions for almost fifty years now, ever since, as a young physics student between semesters at Rice University (during the "Summer of Love" in 1967), I first experienced the full impact of ranges and dimensions of my mind tuning in to bandwidths far beyond my normal spectrum of awareness, beyond anything I could ever have previously imagined.

My lifelong search for a map of consciousness began around a small campfire on a lovely Pacific Ocean beach just south of Big Sur, California, at around midnight on a clear July night, shortly after ingesting three small yellow tablets of LSD ("Owsley acid"). Having just completed the third year of a double major program in mathematics and electrical engineering, my vision of the universe was deeply influenced by such things as frequency charts, Fourier and Laplace transforms, and electromagnetic theories describing invisible energy waves. Suddenly all of these dry theories became vividly alive for me, overlaid and energized by the direct visual, auditory, and tactile experience of other entities in an enormous sea of energy. Astonished, I found myself floating on the vast depths of oceans of consciousness that I had never realized lay beyond the shores of my own familiar mind, and I realized that we are all immersed in planetary and galactic energy fields swirling in and out of our own limited islands of awareness.

The map presented in this book is derived from direct experience. It is also an extension of several major new concepts that have emerged in recent quantum physics and brain research, primarily the "holonomic mind/brain" theory put forth by the neurosurgeon Karl

Pribram and the cosmology of the "Implicate/Explicate orders" proposed by the British quantum physicist David Bohm. Woven together, a synergy emerges as the *holoflux hypothesis*, revealing a new map of consciousness in an expanding universe. The holoflux hypothesis presented here offers a new paradigm which is shown to be consistent with physics and found surprisingly to be in accord with the writings and verbal traditions of the numerous metaphysical interpretations brought back by millenia of introspective mystics, saints, and shamanic psychonauts.

CONTENTS

Chapter 1: The Integral Approach

This book describes a theory of consciousness that is in accordance with the scientific principles of physics and mathematics as well as with the many experiential metaphysical descriptions of mystics and philosophers. The theory is developed through application of an integral methodology that is more than simply transdisciplinary. While it does include the effort to discern patterns and elicit connections from a set of normally non-intersecting academic and scientific disciplines, the integral method fully acknowledges the importance of a participatory, experiential dimension, particularly relevant to any real progress in the study of consciousness. Thus the theories of Karl Pribram and David Bohm are presented sequentially within the context of their unique life experiences.

Integral Philosophy and Integral Theory

With an ever-growing flood of information and data to assimilate, academic specializations have had to grow more focused. However, excessive focus and specialization has resulted in a human knowledge that is becoming increasingly fragmented. The archetypal pattern of the Tower of Babel comes to mind; each domain of knowledge has developed its own highly specialized language and conceptual framework, such that cross-departmental dialog has been impaired. According to the systems thinker, Edgar Morin, the problem seems to be fragmentation and lack of integration:

> An influx of knowledge at the end of the 20th century sheds new light on the situation of human beings in the universe. Parallel progress in cosmology, earth sciences, ecology, biology and prehistory in the 1960s and 1970s have modified our ideas about the universe, the earth, life and humanity itself. But these contributions remain disjointed. That which is human is cut up into pieces of a puzzle that cannot form an image . . . . The new knowledge, for lack of being connected, is neither assimilated nor integrated. There is progress in

knowledge of the parts and paradoxical ignorance of the whole.[1]

The integral methodological approach in this book assumes that valid data may be found significantly beyond the traditional domains of scientific methodology. An integral method examines information from multiple and often disparate domains, seeking to perceive correlations between them, resonances among disparate domains of knowledge that might lead to conceptual bridges between these otherwise relatively isolated, compartmentalized knowledge.

An integral methodology has been explored and fostered in graduate studies programs and research at the California Institute of Integral Studies in San Francisco since the middle of the last century. In 1950, Dr. Haridas Chaudhuri traveled to the United States at the request of the Stanford professor of Asian religions, Frederic Spiegelberg, who had approached the philosopher–sage Sri Aurobindo to recommend an Indian professor to head a center of East–West studies in San Francisco. Chaudhuri left India, where he had been Chair of Philosophy at the University of Calcutta, and eventually founded what is today the California Institute of Integral Studies.[2] In California, Chaudhuri developed his own version of integral philosophy, which he termed "integral panentheism" or "integral nondualism," based upon the writings of Aurobindo, for whom evolutionary energy is to be found simultaneously manifest in the cosmos of space–time as well as in a transcendent region, beyond time or space.[3]

Aurobindo's Integral Perspective

This new approach to research, the integral methodology, stems from earlier efforts in East/West studies which soon evolved into initial efforts at what might be called integral philosophy and integral systems theory.[4]

The Cambridge-educated Indian philosopher, Aravinda Ackroyd Ghose (1872–1950), was possibly the first writer in English to have an integral perspective and to use the word "integral" in his own philosophical writings exploring consciousness and evolution. Later in his life Aravinda was referred to as Sri Aurobindo. "Sri" is a Sanskrit word denoting "that which glows with radiance," and is an honorific title often encountered within Indian culture, reserved for the most highly revered individuals, heavenly beings, or holy scriptures.

Aurobindo left India at the age of 6 and only after completing academic training at King's College, Cambridge, did he returned to India, at age 20. For six years he worked both as a professor and as an administrator for the powerful Maharaja of Baroda in the western coastal state of Gujurat, 800 miles west of Calcutta. Within a few years his growing involvement with other intellectual Bengalis, intent on overthrowing the British Raj and establishing a free sovereign nation, led to his relocation to Calcutta, where he and his friends soon began a fiery newspaper, *Bande Mataram*, or "Victory to the Mother [Mother India]."

Aurobindo was the primary editor, contributing numerous highly articulate but incendiary articles to the newspaper, and became caught up in the organizing and encouragement of several groups of young revolutionaries. This ultimately led to an assassination attempt on a British chief magistrate, Douglas F. Kingsford, who was playing bridge at a British club in Calcutta. Unfortunately, the two bombers ran up to the wrong carriage, which had left the club that night shortly before the one carrying Kingsford. Instead of killing the magistrate, the bomb landed in the carriage of a British mother and her daughter, killing both occupants and the driver. Although no evidence was discovered to directly implicate Aurobindo in the bombing, he was swept up by the police along with a dozen of his colleagues in the independence movement and sent to prison. Several days later he was arraigned at the initial preliminary hearing preceding the trial officially designated as "The Emperor vs. Arabindo Ghose and Others," otherwise known as the Alipore Bomb Trial.[5]

He himself was put in isolation in a stone cell, five feet wide by five feet long, in the Alipore jail south of Calcutta. Such enforced isolation of an individual is not always a setback in the life of a great spirit, however, and in Aurobindo's case we are reminded of the effect of enforced imprisonment upon the lives of such notables as Ghandi and Nelson Mandela.

During the many months of isolation in his cell awaiting trial, Aurobindo began a daily practice of meditation, contemplation, and fasting. Within several months he began to have what he thought to be direct experiences of what he later termed the "silent, spaceless, and timeless Brahman." He resumed his practice of deep breathing exercises, *pranayama*, which he had learned during his days as a professor in Baroda. Here, at last, in the silence and isolation of this punishment cell, he had adequate time to practice *pranayama* for the first time in a sustained manner. At one point after a prolonged fast, he experienced such a flood of inner energy that he felt no need to eat. This was convenient, as the food was somewhat repulsive. It was also during this time that he found he did not require much sleep, and later told a friend that he only needed to sleep one night out of three. During the day he would often meditate while gazing at a large tree outside his jail cell window; one day, to his astonishment, the tree's leaves assumed a bluish cast, and he felt deeply the presence of Sri Krishna (the blue-skinned Hindu God of compassion, tenderness, and love).

Within the same week he suddenly began hearing a voice during his contemplation, which he recognized as the voice of Swami Vivekananda, a Bengali compatriot ten years Aurobindo's senior who had died in 1902. Like Aurobindo, Vivekananda had also completed a bachelor degree in philosophy, but unlike Aurobindo, he had studied directly under the guidance of a mystic, Sri Ramakrishna, a Bengali saint and guru. For almost three weeks, during periods of deep contemplation, Aurobindo listened to the calm, reassuring voice of Vivekananda, of which, he was later to write, "taught me several things about the processes of the higher truth consciousness."[6]

Many years later, Aurobindo, looking back on this period in his life, comments on his first day in prison:

> Friday, May 1, 1908 . . . I did not know that day would mean an end of a chapter in my life, and that there stretched before me a year's imprisonment during which period all my human

> relations would cease, that for a whole year I would have to live, beyond the pale of society, like an animal in a cage. And when I would re-enter the world of activity, it would not be the old familiar Aurobindo Ghose. . . . I have spoken of a year's imprisonment. It would have been more appropriate to speak of a year's living in an ashram or a hermitage. . . . The only result of the wrath of the British Government was that I found God.[7]

Within several months following his release from prison, Aurobindo received word that various agencies of the British Raj were about to reopen charges against him and that they did so intending to deport him to a British penal colony from which few ever returned, located in the Andaman Islands, 800 miles southeast across the Bay of Bengal near the coast of Thailand. Immediately Aurobindo made plans to depart from the British territories to maintain his freedom. With the help of his closest friends, he fled by a circuitous sea route, finally arriving at the port city of the small French colony of Pondicherry, on the Indian coast about a hundred miles north of Jaffna, Sri Lanka. In Pondicherry, to the surprise and concern of many of his close Bengali political associates, Aurobindo dropped all involvement in the politics of independence. Instead, he spent the next four years reading philosophy and living the life of a reclusive young contemplative.[8] Having become fluent in the French language during his years at Cambridge, it is likely that during his Pondicherry years Aurobindo became acquainted with the writings of Henri Bergson, who in 1911 had published *Creative Evolution*.[9]

In *Creative Evolution*, Bergson argues for "intuition" as being of primary import in advancing any really new understanding of consciousness. He sees intuition as being clearly a mode of conscious awareness *beyond* the domain of an intellectual consciousness, and goes on to criticize modern epistemologies for discounting the efficacy of intuition while enforcing the primacy of intellect:

> In the humanity of which we are a part, intuition is, in fact, almost completely sacrificed to intellect . . . . Intuition is there, however, but vague and above all discontinuous. It is a lamp almost extinguished, which only glimmers now and then, for a few moments at most . . . feeble and vacillating,

but which none the less pierces the darkness of the night in which the intellect leaves us.[10]

After four years of study, contemplation, and silence, Aurobindo once more began writing, and soon published, on a monthly basis, a wide-ranging collection of philosophical essays. In these essays, Aurobindo can be seen expressing himself from the vantage point of an *integral perspective* in the broadest sense. His mature writings result from an integration of his knowledge of Western evolutionary ideas based upon his Cambridge education, overlaid upon his intuitive, experiential observations of consciousness obtained through direct contemplative experience.[11]

Aurobindo is often credited as the first writer to use the term "integral" in connection with evolutionary philosophy, however, Harvard sociologist Pitirim Sorokin also began using the phrases "integral philosophy" and "integralist" in this connection at around the same time. In fact, it appears that Sorokin, Aurobindo, and Gebser each originally adopted the term "integral" independently without knowledge of its use by others.[12]

Gebser's Integral Perspective

A contemporary of Aurobindo, the German cultural historian and evolutionary philosopher Jean Gebser, writing in Switzerland in the 1940s, developed another approach to integral philosophy. Laying out his theory of integral consciousness, in *The Ever-Present Origin*, Gebser develops a theory of the evolutionary emergence of what he calls "the 'aperspectival (arational–integral)' consciousness." Gebser's motivating objective was to foster what he called an incipient "integral aperspectival consciousness."[13]

Gebser's own research focus was primarily within cultural anthropology, but his integral methodology is far ranging. In it can be seen a search for meaning through the discovery of transdisciplinary connections. In an essay entitled "Cultural Philosophy as Method and Venture," Gebser explains the importance of an integral effort for uncovering "meaning giving connections:"

> It is a scientific attempt, particularly distinctive of our epoch, to appraise the multiplicity of cultural endeavors, that is to say, to uncover meaning-giving connections . . . it could contribute toward the abolition of the feeling of isolation which today not only reigns in the various scientific disciplines but which is generally prevalent.[14]

The German philosopher Georg Feuerstein, in discussing Gebser's methodology, describes how Gebser's aim can be seen encapsulated in the term *synairesis*, "an intelligent realization of reality proceeding from the accomplished transparency of all modes of cognition."[15] In fact, Feuerstein tells us, it is this *synairesis* of Gebser's that can free us from the limited bounds of scientific realism. Feuerstein here comments:

> Gebser also makes it clear that his panoramic orientation is not a mere synthesis, which would only be part of the kind of systematization peculiar to the mental-rational consciousness. He speaks of his methodology as aiming at *synairesis*, which is an integral understanding, or perception, of reality . . . . *Synairesis*, then, describes an intelligent realization of reality proceeding from the accomplished transparency of all modes of cognition.[16]

Gebser coins another term to describe his methodology, *systasis*, a term implying a process whereby partials merge or are merged into a whole:

> *Systasis* is the means whereby we are able to open up our consolidated spatial consciousness to the integrating consciousness of the whole. This integrating consciousness enables us to perceive and present the integrity or integrality of the whole.[17]

By contrast, scientific materialism, the driving operational paradigm behind contemporary science and technology, includes what is clearly a limited domain of inquiry in its definition of reality. This limitation is gradually being recognized by small segments of the scientific community, as Feurstein remarks here:

> Only a dimension that is quantifiable, measurable, analyzable, abstractable, or predictable . . . one that is flat and monochrome, and needless to say, is contradicted by the

> scientific evidence itself. It is for this reason that avant-garde scientists admit the need and indeed are looking for a new "paradigm" that, hopefully, will do justice to the multidimensionality of humanness.[18]

Gebser's vision points the way to a new research methodology, an integral, multi-perspectival approach. From a wider mix of perspectives than traditionally afforded by academic analysis, the focus here is comprised of discovering a pattern, a *synairesis* emerging from multiple disciplines. From this multi-domained perspective new connections are discoverable from which new understandings may emerge. Figure 1 presents a visual diagram of the integral methodology applied specifically in this book, in which can be seen the hard problem of consciousness approached from the following eight perspectives:

- Consciousness Studies
- Electrical Engineering
- Dimensional Analysis
- Speculative Philosophy
- Quantum Theory
- Mathematics
- Geophysics
- Human Physiology

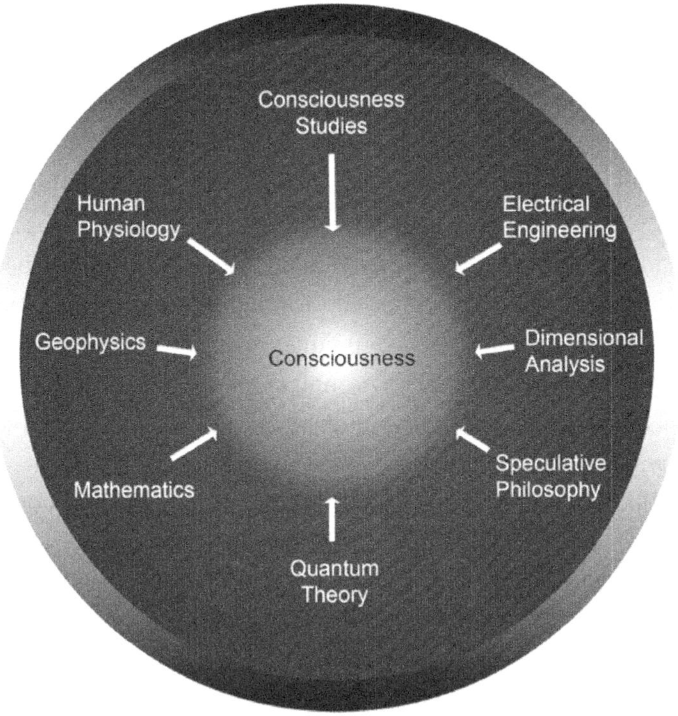

Figure 1. *An integral focus on consciousness.*

Wilber's Integral Perspective

A third and equally elucidating perspective on *integral philosophy and methodology* is to be found in the writing of American philosopher-psychologist-integralist Ken Wilber, who describes his own vision of an integral perspective and method as follows:

> If we look at the various fields of human knowledge—from physics to biology to psychology, sociology, theology and religion—certain broad, general themes emerge about which there is very little disagreement. If we take these types of largely agreed upon orienting generalizations from the various branches of knowledge … and if we string these orienting generalizations together, we will arrive at some astonishing and often profound conclusions, conclusions that, as extraordinary as they might be, nonetheless embody nothing more than our already agreed upon knowledge.[19]

The strength of Wilber's contribution is that he has been able to integrate material from far ranging disciplines, including psychology, Asian philosophy, and to a lesser extent, the sciences, and he has developed a methodology which he terms Integral Methodological Pluralism (IMP), in which multiple perspectives are brought together in an effort to elicit a deeper understanding of consciousness.[20] The methodology of IMP stresses the idea that all articulated understandings are context based, formed within specific cultural contexts, and molded by the unique academic, experiential, and cultural-historic background environments from which each individual attempts to formulate their own understanding and with which to give their own unique expression.[21]

Wilber's methodology often works by juxtaposing older historical paradigmatical frameworks alongside newer paradigms. In his 1996 *A Brief History of Everything*, for example, in developing a point he might discuss contemplative maps of consciousness developed by Indian Vedantic traditions, and then contrast these with quantum mechanical considerations from physics, and finally analyze both in terms of various transpersonal theories expressed by modern psychologists.

Wilber also describes the recent emergence of an evolutionary epistemological capacity which he calls "vision-logic." Wilber's vision-logic consists of an integration of intellectual capacity with intuition in order to bring about the recognition of new relationships among seemingly radically diverse perspectives, with the objective being to discover congruencies that might bring forth new understandings of the phenomena being studied.[22]

Wilber's vision-logic, with its emphasis on the application of intuition, echoes the central methodological idea of "intuition" as developed by Henri Bergson, but Wilber's approach is also strongly influenced by the integral methodology developed by Jean Gebser.[23]

In summary, this book, following the approaches of Aurobindo, Bergson, Gebser, and Wilber, is an effort to bridge the gap between the natural sciences and philosophical speculation. Relevant observations from various natural sciences and humanistic disciplines are examined, compared, and coordinated to support a holoflux field theory of consciousness. Nevertheless, science (and in particular mathematics and physics) continues to be key to the articulation of a workable theory of consciousness. In a recent essay, "How Consciousness Becomes the Physical Universe," Kafatos, Tanzi, and Chopra declare:

> Physics can serve a pivotal role in transitioning to this new model, because the entire biosphere operates under the same generalized principles we described from the quantum perspective, as does the universe itself. This simple unifying approach must be taken, we realize, as a basic ontological assumption, since it cannot be proven in an objective sense. We cannot extract consciousness from the physical universe, despite the fervent hope of materialists and reductionists . . . . Our approach, positing consciousness as more fundamental than anything physical, is the most reasonable alternative: trying to account for mind as arising from physical systems in the end leads (at best) to a claim that mathematics is the underlying "stuff" of the universe (or many universes, if you are of that persuasion.) . . . . In that sense, we believe the ontology of science will be undivided wholeness at every level. Rather than addressing consciousness from the outside and trying to devise a theory of everything on that basis, a successful Theory of Everything (TOE) will emerge by taking wholeness as the starting point and fitting the parts into it rather than vice versa.[24]

Information, Meaning, Experience

As a result of difficulties in identifying a consensual strategy to exploring consciousness, the scientific community seems to have sidestepped the entire project, for it appears that, as the Australian philosopher David Chalmers has observed, "existing theories of consciousness either deny the phenomenon, explain something else, or elevate the problem to an eternal mystery."[25] Chalmers himself expresses a view that consciousness ought to be approached through a search for a fundamental link between information and experience, and goes so far as to conjecture that "phenomenal properties are the internal aspect of information."[26]

Information and meaning are also central to the concern of the quantum physicist David Bohm, who sees that, "Meaning is inseparably connected with information."[27] For Bohm the search for meaning is seen as an unfolding of fresh perceptions of reality, an unfolding which occurs when sustained attention and sharp intentional focus is brought to bear upon a specific problem:

> When we use the term "meaning," this includes significance, purpose, intention and value. However, these are only points of departure into the exploration of the meaning of meaning. Evidently, we cannot hope to do this in a few sentences. Rather, it has to be unfolded as we go along. What is needed is a creative attitude to the whole, allowing for the constantly fresh perception of reality, which requires the unending creation of new meanings. This is especially significant, in the exploration of the meaning of meaning.[28]

Bohm, in what he calls the "causal interpretation" or "ontological interpretation" of quantum field theory, tells us that *meaning unfolds* from out of an implicate domain, a dimension transcendent of space and time. This unfolding meaning is expressed in space–time, which Bohm calls the explicate order. Bohm's model is one that can be described by the "quantum potential" function, Q, which Bohm and Peat derived from Schrödinger's wavefunction (which describes all moving particles as wave-like functions).[29] Bohm's theories are discussed at length in Chapter 4 of this book.

But perhaps the most widely debated issue in consciousness studies can be found encapsulated in the phrase "the hard problem of consciousness," first coined by Chalmers in a 1995 essay, in which he places emphasis squarely upon the issue of *experience*.

Chalmers' Hard Problem

Chalmers proposes a theory of consciousness which highlights the importance of information, while conflating it into what he calls, the "hard problem":

> The really hard problem of consciousness is the problem of *experience*. When we think and perceive, there is a whir of information processing, but there is also a subjective

> aspect . . . . Why should physical processing give rise to a rich inner life at all? It seems objectively unreasonable that it should, and yet it does. If any problem qualifies as *the* problem of consciousness, it is this one.[30]

This "whir of information processing" Chalmers refers to is readily observed in the electrical information processing of the neuronal brain system in the body, like the "thinking meat" which so astounded two alien observers in a story by the science fiction writer, Terry Bisson:

> "They're meat all the way through."
>
> —"No brain?"
>
> "Oh, there is a brain all right.
> It's just that the brain is made out of meat!"
>
> —"So . . . what does the thinking?"
>
> "You're not understanding, are you?
> The brain does the thinking. The meat."
>
> —"Thinking meat!
> You're asking me to believe in thinking meat!"
>
> "Yes, thinking meat! Conscious meat! Loving meat.
> Dreaming meat. The meat is the whole deal!
> Are you getting the picture?"
>
> —"Omigod. You're serious then.
> *They're made out of meat!*"[31]

The data stored and manipulated within modern electronic technologies is a property of physical processes, and only becomes information when presented to the conscious perception of a human reader. Within the computational space of the computer itself, the information is merely symbolic code, assumed physically to be unconscious. Chalmers posits a theory of consciousness that is the union of experience and the physical:

> A non-reductive theory of consciousness will consist in a number of *psychophysical principles*, principles that connect the properties of physical processes to the properties of experience.[32]

Chalmers declares that the basic principle of consciousness must involve the notion of *information.* He discusses how physically embodied *information states* must be embedded in an *information space* before it can become active and accordingly, transmittable:

> An information space is an abstract object, but following Shannon we can see information as *physically embodied* when there is a space of distinct physical states, the differences between which can be transmitted down some causal pathway. The transmittable states can be seen as themselves constituting an information space. To borrow a phrase from Bateson (1972), physical information is a *difference which makes a difference.*[33]

Information then, to Chalmers, has two basic aspects, a phenomenal aspect and a physical aspect, yet he questions whether all information has a phenomenal aspect. If there is no such constraint, he speculates, then experience might be seen as being a fundamental property of the cosmos. If there is no phenomenal constraint, Chalmers says,

> . . .then experience is much more widespread than we might have believed, as information is everywhere. This is counterintuitive at first, but on reflection the position gains a certain plausibility and elegance . . . . Indeed, if experience is truly a fundamental property, it would be surprising for it to arise only every now and then; most fundamental properties are more evenly spread . . . . Once a fundamental link between information and experience is on the table, the door is opened to some grander metaphysical speculation concerning the nature of the world . . . . The theory I have presented is speculative, but it is a candidate theory.[34]

Chalmers favorably compares his theory to that put forth by John Archibald Wheeler (1911–2008), the quantum physicist who coined the term "Black Hole."[35] Wheeler, who received his PhD in physics at the age of 21, suggested that information is fundamental to the physics of the universe, and regularly proclaimed that he regarded "this idea of information theory as the basis of existence.[36] Wheeler claimed that he could summarize his understanding of the cosmos with the catchphrase "It from Bit," and explains how it implies that information may be the basis of the universe:

> It is not unreasonable to imagine that information sits at the core of physics, just as it sits at the core of a computer. Trying to wrap my brain around this idea of information theory as the basis of existence, I came up with the phrase "it from bit."[37]

Wheeler and one of his graduate students went on to calculate the information storage capacity on the spherical boundary of a black hole, and came up with an estimation of astronomical storage capacity, according to which a Bekensteinian sphere with the diameter of a typical human blood cell would have the storage capacity of 10^{60} information bits.[38]

Electromagnetic Fields as Mind

Several twenty-first-century researchers in the life sciences have concluded that consciousness might have a basis in electromagnetic field phenomena. In 2000, a neurophysiologist in New Zealand, Susan Pockett, published a paper hypothesizing that consciousness may very well be the electromagnetic field itself. Pockett states:

> Perhaps what we have been looking for is that what has in the past been called the electromagnetic field is itself consciousness. And perhaps our individual consciousness, which unlike the field as a whole *are* bounded in space and time, are identical with particular local spatiotemporal configurations of the electromagnetic field.[39]

Elsewhere, she states that, "The essence of the hypothesis was that conscious experience (a.k.a. sensation) would prove to be identical with certain spatiotemporal patterns in the electromagnetic field,"[40] and indicates that perhaps researchers have been looking for consciousness in the wrong ranges of the electromagnetic frequency spectrum:

> The signals we are looking for in brain-generated electromagnetic activity might actually NOT be in the ELF (extremely low frequency) range where everyone is presently looking, but could turn out to be closer to the microwave frequency range which is of course always filtered out in present day electrophysiological recordings.[41]

It is important to note that Pockett assumes that consciousness, whatever the mechanism or bandwidth, is a recent epiphenomenon of some evolutionary process, a serendipitous emergence. This can be seen in her concluding remarks, "The present theory is quite compatible with the view that consciousness was not present in the universe before its biological evolution."[42]

In 2002, Johnjoe McFadden, a professor of molecular genetics at the University of Surrey, put forth a conscious electromagnetic information (CEMI) field theory.[43] McFadden hypothesizes that an electromagnetic information field of the brain is the underlying basis of consciousness. In 2007, McFadden identifies *information* as the agential binding element between the physical and the conscious:

> The key feature of the brain's electromagnetic field is that it is capable of integrating vast quantities of information into a single physical system and it thereby accounts for the binding of consciousness. Unlike quantum theories of consciousness, the CEMI field theory does not require any special physical states in the brains; it is perfectly compatible with brain physiology. Nevertheless, recent work has shown that classical electromagnetic waves may be used to implement quantum algorithms; therefore the brain's CEMI field may be able to perform quantum computations (but without the requirement for quantum coherent states of matter).[44]

Like Pockett, McFadden categorically links consciousness to the electromagnetic field: "I propose that the brain's electromagnetic information field is the physical substrate of conscious awareness."[45] Yet whatever the degree of enthusiasm Pockett and McFadden have for the electromagnetic conjecture, their theory ends there, with no correlative suggestion for further research. The work of the neurosurgeon Karl Pribram, on the other hand, in searching for a neurophysical basis for memory, also embraced an electromagnetic field theory of consciousness. During a lifetime of research, he was able to formulate and publish a coherent, evidence-based, holonomic brain theory based upon a mathematical Fourier-like transmission and transformation of information.[46]

Chapter 2: Karl Pribram

The electromagnetic holoflux theory of consciousness rests upon a series of basic concepts, and in particular the holonomic brain theory developed by the Stanford professor emeritus Karl Pribram (1919–2015). Pribram's neurosurgical and mathematical search for a reasonable solution to the so-called "hard-problem" of consciousness advanced steadily during his 70-year career of experimental research. His breakthrough metaphor, reinforced through subsequent experimental-driven research, came in the form of a hologram:

> Using the mathematical holographic process itself as a metaphor . . . my (1991) book *Brain and Perception* provides detailed review of experimental results that support the conjecture that holography is a useful metaphor in coming to understand the brain/mind relation.[47]

Pribram, who has been called "the Magellan of the brain," published in his lifetime over 700 papers and 20 books on pattern perception, cognition, and the brain. In 1972 he put forth a "holonomic brain theory," a mathematical model of brain function in which neurodynamics aligns with quantum theory.[48] Pribram here explains the use of the term "holonomic" in his theory:

> The term for the theory, *holonomic*, was first used by Hertz to describe linear transformations when they are extended into a more encompassing domain. I have here extended its meaning to cover the spectral domain.[49]

The *spectral domain* to which Pribram here refers is also known as the *frequency domain* in quantum physics and electrical engineering. The spectral domain as a concept has been developed further in the mathematics of the Fourier transform whereby information signals in a space–time domain are transformed into a complex of frequencies in a spectral domain.[50]

Early in his career, Pribram had become fascinated by the search for the engram, the suspected location of memory recordings in the brain.[51] In the 1920s the Canadian neurosurgeon Roger Penfield, while operating on the exposed cerebral cortex of patients with severe epilepsy, discovered that electrical stimulation of small areas of the parietal lobe would cause the patients, who were always awake during such operations, to recall specific memories, often in vivid detail.[52] According to Penfield:

> It was evident at once that these were not dreams. They were electrical activations of the sequential record of consciousness, a record that had been laid down during the patient's earlier experience. The patient "re-lived" all that he had been aware of in the earlier period of time as in a moving-picture "flashback."[53]

This discovery provided evidence to support the existence of the *engram*, a hypothesized physical storage location of memory traces, but whose mechanism and location has never been determined, in spite of decades of persistent research.[54]

Neuroanatomical Concepts

Before considering Pribram's holonomic brain theory in detail, a brief review of neuroanatomical terms and concepts here provide a context for a critical examination of Pribram's research and theories. A typical neuron cell (Fig. 2) consists of a single nucleus, a single axon, and multiple dendrites, as can be seen in the image in the left of the figure.

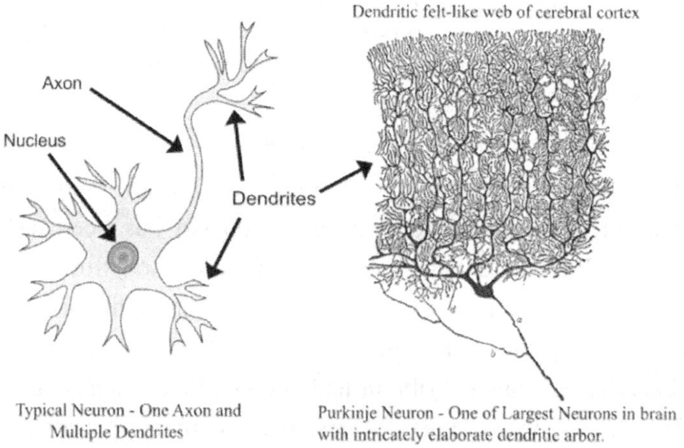

Figure 2. *Typical neuron and Perkinje neuron.*[55]

All cells have a nucleus, but the neuron has two types of neuronal fiber extrusions: a single, usually relatively long axon, long in comparison to the dendrite branches, which are many and complexly bifurcated.[56] The longest axons in the body belong to the sciatica nerves that run from the brain down through the spine and ending in the big toe. Axons are also known as "nerve fibers" and most neurological research focuses on these fibers as they carry the relatively large "nerve impulse" electrical spikes that are easily measured. The numerous dendrites, on the other hand, resemble finely bifurcating tree-like branches that gather in highly complex felt-like regions.

The thick-layered dendritic region of Perkinje neurons to the right in Fig.2 can be seen in greater detail as they manifest in various layers in Fig. 3 where a typical section of the cerebral cortex is depicted. Surprisingly thin, the entire cerebral cortex, at its thickest, is only 2 millimeters in depth, about the thickness of four human hairs placed side-by-side.[57] Three aspects of a 2mm-thick cerebral cortex slice can be seen in Fig. 3: in the left area, a single sensory nerve fiber is shown; in the center of the drawing is revealed an interlinked systems of fibers, grouped into four cell-type layers as labeled on the left axis; finally, the right-most region of the figure indicates fiber bundle types.[58]

The cerebral cortex, in spite of being extremely thin, occupies fully 88% of the human brain mass.[59] This is due to its extremely intricate three-dimensional folding and refolding in what appears to be a fractal pattern. Within this cortex volume can be found three distinct types of cells: neurons, glial cells, and blood capillary cells, and as will soon be discussed, the omipresent capillary duct system not only provide nourishment for all other cells, as well as removing waste products, but offers the physically contiguous network that would provide for a charged plasma flow that may have much more to do with what we call consciousness than do the electrical spikes of neurons.

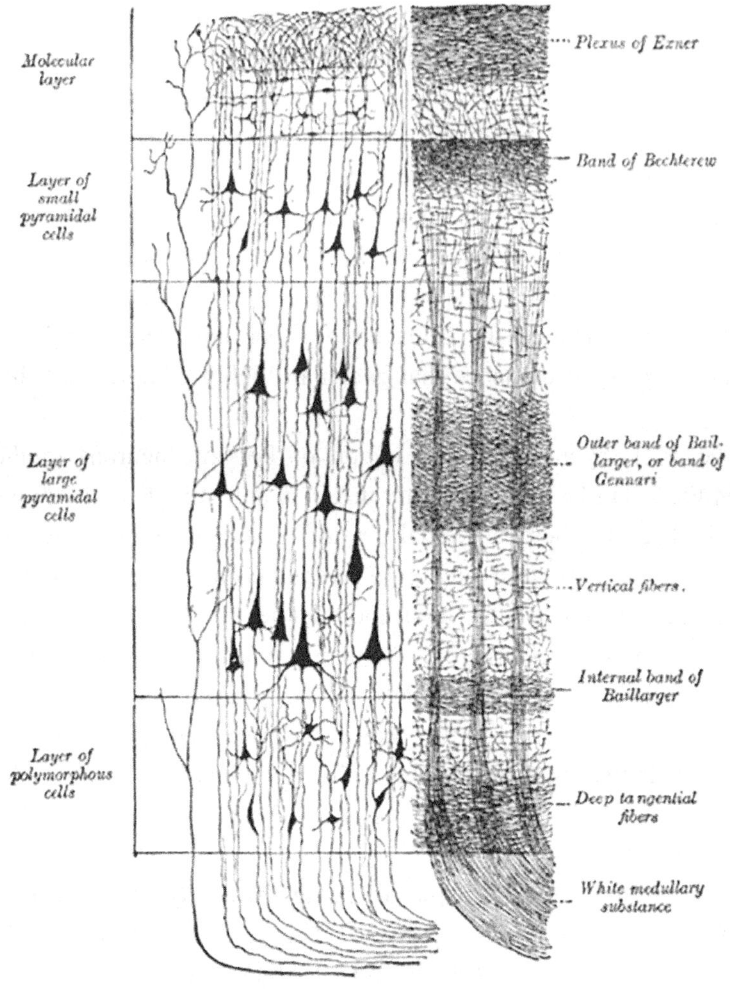

Figure 3. *Cerebral cortex.*[60]

While the major focus of scientific research is upon the 86 billion neuron cells in the human brain, glial cells and capillary cells have been generally dismissed as having no role in cognition or consciousness. Glial cells, since their discovery in 1856, have been tacitly dismissed in the widespread agreement that they have evolved primarily for mechanical purposes, acting in supportive, structural roles (the word *glia* in Greek γλοία means "glue"). Glial cells do not conduct electrical impulses, as do neurons (even though glial cells wrap around neuron axons to form myelin sheaths and thus are capable of modulating the speed and strength in the neural conduction of electrical impulses passing through these glial "wraps"). Likewise, any possible function of the blood capillary cell system in consciousness has also been dismissed; it is assumed that the function of the circulatory system is fully understood, and not involved in thought or conscious sensation.[61]

Memory: The Search for the Engram

Upon graduation with an MD in Neurosurgery from the University of Chicago at the age of 21, Pribram found employment as an assistant to the renowned zoologist Karl Spencer Lashley, who at the time was the head of the Yerkes Laboratory, a joint Harvard/Yale research facility. There, Pribram assisted Lashely and his staff in conducting memory research on the brains of primates.[62]

Lashley had been searching for the location of the elusive memory engram for over 30 years, but during his work with Pribram, the two scientists reached the surprising conclusion that the engram did *not* exist in the cortex at all. They based their conclusion on the fact that no matter how much of a rat's cerebral cortex was removed, the rat would continue to remember its route through a maze that had been previously learned over a period of several weeks.[63] Lashley stated his conundrum as follows:

> Here is the dilemma. Nerve impulses are transmitted over definite, restricted paths in the sensory and motor nerves and in the central nervous system from cell to cell, through definite intercellular connections. Yet all behavior seems to be determined by masses of excitation, by the form or relations or proportions of excitation within general fields of activity, without regard to particular nerve cells. It is the pattern and not the element that counts. What sort of nervous organization might be capable of responding to a pattern of excitation without limited, specialized paths of conduction? The problem is almost universal in the activities of the nervous system and some hypothesis is needed to direct further research.[64]

The search for such a hypothesis fired Pribram's research interest, and, working with Lashley, Pribram discovered areas of the brain surface that, when stimulated or removed, significantly affected the processing of sensory information, even though no direct sensory links existed between neurons in the sensory organs and the areas in question. Pribram called these areas of the cerebral cortex "intrinsic," having no direct connection to the sensory organs, and contrasted these with "extrinsic" areas of the cerebral cortex in which could be traced identifiable, direct neuronal connections to sensory organs.[65]

In 1948, Pribram left Florida for New England to accept an offer to join the faculty at Yale, where he had found the Gestalt theorist Wolfgang Köhler's pioneering dissertation in the field of psychoacoustics to be of great interest.[66] Köhler was famous at Yale for having emigrated from Nazi Germany in 1935, after refusing to acquiesce, as a full professor, to giving the Nazi salute at the beginning of each class. Pribram was attracted to Köhler's approach to research, which had been formed by the Gestalt paradigm "that the whole is *different* from the sum of the parts."[67]

Pribram soon began working on an auditory cognition project at a laboratory in Massachusetts run by Köhler, whose ideas reinforced Pribram's own; if the whole is discovered to be something much more than simply the sum of the parts of which it is comprised, then perception and cognition, in this case the psychoacoustic perception that interested Köhler, might indeed be something categorically different from that of the functioning of sensory neurons capturing light and sound.[68]

Pribram did not share Köhler's view of perception, a view commonly held among brain researchers at the time, that somewhere on the cerebral cortex would be found direct current electrical analogues, two dimensional voltage images perhaps, having the same physical shape as the objects perceived; such a view, held in common by the Gestalt psychologists, is often termed "geometric isomorphism," (*iso* = same, *morph* = shape).[69] Pribram doubted Köhler's supposition that direct current fields would eventually be found to support the theory of geometric isomorphism, but agreed that electromagnetic fields were somehow involved in visual, auditory, and cognitive perception, and in particular, something Pribram termed "minding." As he understood the term, "Minding is attending, looking out and ahead, grounded on looking back and within."[70]

For over two years Köhler and Pribram made extensive measurements of both ape and human subject brains, trying to detect voltage changes between locations on the cerebral cortex surface which would correspond to changes in sensory stimulation.[71] Pribram's procedure was fairly simple, though it required access to the surface of the cerebral cortex, accessible immediately beneath the cranial bones.[72]

During the eight years Pribram worked with Köhler, a typical series of experimental procedures might involve holding up alternately white and then black cards while measuring any detectable and repeatable voltages changes between two adjacent surface areas of the cerebrum; to support his theory, Köhler sought to locate patterns of voltage gradient on the cerebral cortex itself that would be a mirror of the optical visual pattern falling upon the optic field receptor neurons of the eye.[73]

Köhler's Theory of Geometric Isomorphism

After several years of research in Köhler's laboratory, Pribram, who had never believed in geometric isomorphism, designed a successful experiment that provided a refutation of this assumption. Shortly thereafter, an amusing exchange occurred in which Köhler asked Pribram, "Now that you have disproved not only my theory of cortical function in perception, but everyone else's as well, what are you going to do?" to which Pribram replied, "I'll keep my mouth shut."[74]

In the mid-1950s Pribram's research led him to focus upon the neuro-electric activity associated with the fine-fibered webs formed by dendrites, rather than the more obvious electrical impulses which are seen to run along the much thicker axon trunks:

> At about the same time, in the 1950s, that Hodgkin had begun his work on the retina, I shared a panel at a conference held in 1954 at MIT with Stephen Kuffler of the Johns Hopkins

> University . . . . Kuffler focused my thinking on the type of processing going on in the fine-fibered webs of the nervous system, a type of processing that has been largely ignored by neuro-scientists, psychologists and philosophers.[75]

Based upon his half century of research as a neurosurgeon, Pribram described how, it seemed to him, the thick, richly intertwining fine-fibered dendrite branch regions are where focused islands of complex frequency activity are to be found:

> The fine fibers composing the web—the branches or dendrites of nerve cells—cannot sustain nerve impulse conduction the way axons can. The electrical activity of dendrites rarely "sparks" but oscillates between excitation and inhibition; thus, dendritic activity occurs locally, forming patches or nodes of synchronized oscillating activity . . . . When the formative processing web provided by the dendrites has to be accessed by the rest of the brain, axons relay the results of these interactions to other parts of the brain or to receptors and effectors in the body.[76]

This "form of the fine-fibered distributed web of the brain" became Pribram's passion and eventually led to his conclusion that the form of the ultra deep structure of thinking is a holographic-like distributed process:

> According to the reasoning proposed . . . the potential of the ultra-deep structure of thinking is realized when the holographic distributed process becomes transformed by way of the memory-motive and memory-emotive processes into cortically engendered complex verbal and nonverbal language and language-like cultural forms.[77]

In 1958, after 10 years at Yale, Pribram accepted an offer from Stanford to continue his research in California as Director of the Center for Advanced Studies. Over the next 30 years, Pribram published over two hundred papers mapping innumerable features of neuroanatomy and brain function.[78] The general consensus among neurophysiologists assumes that perception is a one-way street, that neuronal receptors in the external sense organs transmit information to the brain via neuronal impulses. However during his years at Stanford, Pribram began to find evidence that the process is more of a "two-way" cybernetic transform, with information flowing both ways, simultaneously:

> Critical to understanding is the acceptance of evidence that brain perceptual systems operate as top-down as well as bottom-up processors. It is this evidence that my colleagues and I have spent almost a half-century in amassing. Some 1,500 nonhuman primates, 50 graduate students, and an equal number of postdoctoral fellows have participated. The results of these researches have cast doubt on viewing brain perceptual processing as elementaristic, bottom-up, reflex-arc, stimulus-response—views that still characterize many texts in neurophysiology, psychology, and perception.[79]

Pribram would later put forth the critique that it is this very "neglect of a science based on wider patterns that has resulted in a science based on matter, and an overarching materialism."[80]

During the course of his research, Pribram gathered sufficient evidence to convince him that not only visual reception of neurons seemed to exhibit top-down patterns of integrated field activity, but that auditory and motor systems in the brain also exhibited distributed field patterns of activity and response. As early as 1962, in a paper published on the work of Sigmund Freud, Pribram began to express his interest in focusing on the *patterns* of brain activity, rather than tracing specific neuroanatomical circuit connections:

> For Freud, a neuron may "fill"—i.e., become *cathected*—with excitation even though no transmitted activity results . . . . This emphasis on cathexis is one of those strokes of luck or genius which in retrospect appears uncanny, for only in the past decade have neurophysiologists recognized the importance of the graded non-impulsive activities of neural tissue—graded mechanisms such as those of dendritic networks whose functions are considerably different from those of the transmitted impulsive activity of axons.[81]

The Holographic Paradigm

In the summer of 1965, Pribram first read of *holography* in a *Scientific American* article. He was immediately captivated by the idea of using Fourier transform mathematics and laser radiation to produce visual three-dimensional imagery, and realized that this application of the Fourier transform might offer new approaches to storing enormous amounts of information as raw data. The visible projection of these three-dimensional holograms was a direct result of recent developments in laser technology, a capability that had not been available to the original discoverer of the holographic principle in 1947.[82]

It was at that time that the mathematics of the holographic principle had been pioneered by Dennis Gabor (1900–1979), a physicist and electrical engineer who used the Fourier transform to develop a novel optical approach for improving the resolution of the electron microscope. Gabor had earned his PhD in Berlin by analyzing the properties of high voltage transmission lines through the use of the cathode-ray oscillographs, leading to his interest in electron optics. In 1933 he left Nazi Germany for London, recruited by Thomson-Houston Electric Company (the British subsidiary of General Electric). It was there, in 1947, that Gabor developed the mathematical basis of his holographic method. In referring to his predicted phenomenon, Gabor invented the name *holography* (from the Greek terms *holos*, meaning "entire" or "whole," and *grafe*, "writing").[83]

However, Gabor did not produce an actual three-dimensional hologram, which had to await the future invention of the perfectly coherent light source which would eventually be provided by the not-yet-existent laser technology. Instead, in 1946, Gabor used a highly-filtered mercury arc light to apply his new holographic method to improve the resolution power of electron micrographic microscopes. Unfortunately for Gabor, his actual implementation of the holographic idea was not widely embraced. Although the technique was seen to be theoretically intriguing, and a somewhat esoteric possibility for microscopy, it achieved neither popular acceptance nor financial success due to the time-consuming method required to produce an image.[84]

> Instead of the immediacy of seeing an image on a fluorescent screen (as some electron microscopes then produced), the

reconstructed image was to be obtained more painstakingly via a half-hour exposure, followed by conventional photographic processing, unintuitive optical transformation, and observation through a conventional microscope eyepiece . . . Gabor blamed lack of enthusiasm from his industrial collaborators and microscope manufacturers, and microscopists dismissed what they saw as a hybrid and unfamiliar technique.[85]

In 1958, Yuri Denisyuk, a Russian engineer, working at the Vavilov State Optical Institute in Leningrad, developed a holographic imaging system similar to Gabor's, also using filtered mercury arc lighting. However, as Gabor had experienced in Britain, Denisyuk found little support in the Soviet Union for his holographic process.[86] In 1962, however, Emmett Leith and Juris Upatnieks, working in a highly classified laboratory near the campus at the University of Michigan were trying to develop a new form of side-looking optical radar. In their laboratory they configured an experiment to create a holographic image using the coherent source of light from a laser (a technology itself developed only in 1960).[87] Through their use of a laser light beam they were able to create a three-dimensional holographic image that could be viewed directly, without the requirement of peering through an optical device, such as a stereoscopic microscope. Their significant discovery, as well as the almost identical, simultaneous discovery of the holographic method in the Soviet Union, remained classified for almost two years before the news was revealed to the public.[88]

The creation of a hologram requires two fixed sources of radiation of the same frequency, usually created in practice by splitting one original beam of electromagnetic laser radiation into two separate paths. The two resulting beams of energy (often infrared) are then aimed, from different angles of incidence, to fall upon a three-dimensional object. The two impinging light beams cast complex interference shadows upon a two-dimensional film upon which the patterns are recorded. These shadows form what are termed "holographic interference patterns." In a similar way, the three-dimensional information about the image, captured as complex shadows upon the recording film, can be subsequently used to reproduced the image by reversing the process.[89]

The full sequence of transformation is as follows: the three-dimensional information is first transformed into two-dimensional interference patterns of waves in space–time. The three-dimensional information has been transformed *out of* the space domain *into* a frequency-phase data which can be stored on a flat two-dimensional film. From film, it can be transformed back again into the space domain through the focus of light energy of a single frequency through the complex fringe patterns imprinted on the film. This results in the formation of interference of light waves in space on the other side of the film from the source, phase interferences in the light create a virtual three-dimensional image, identical to the one existing during the original recording. It is this resulting transmitted image that is termed a *hologram*.[90]

A major breakthrough in his search for a nonlinear frequency-based model occurred in 1965 when Pribram learned of the holographic method of memory storage in a *Scientific American* article written by the research partners Leith and Upatnieks.[91] What most impressed Pribram was that the article "described how the recording of interference patterns on photographic film had tremendously enhanced the storage capacity and retrieval of images."[92] The potential for rich data storage in holographic image technology intrigued Pribram, as it seemed to indicate new possibilities for visual processing and a new mechanism for memory storage.[93]

Gabor had been awarded the Nobel Prize in Physics in 1971 for his early work in the mathematics of communications, which eventually led to the invention of the hologram. In his Nobel Lecture acceptance speech, likely much to Pribram's delight, he described the possible relationship of the hologram to human memory:

> In a hologram the information is spread over the entire hologram area . . . . The appearance of such a "diffused" hologram is extraordinary; it looks like noise . . . . A diffuse hologram is therefore a distributed memory, and this has evoked much speculation whether human memory is not perhaps, as it were, holographic, because it is well known that

> a good part of the brain can be destroyed without wiping out every trace of a memory.[94]

That same year, 1971, Pribram published *Languages of the Brain*, summarizing the results of six years of experimental work at Stanford, and supporting a new theory proposing cognitive holographic patterns in the brain. He proposed that while memories might not be localized in any specific brain cell, laboratory data indicates that memories are found to be spread, holographically, throughout wide regions of the fine-fibered web of the brain's dendritic synapses, like complex nets or networks.[95] Pribram's new hypothesis integrates the mathematical paradigm of optical holography with his observations that the processing of parallel information intercepted by sensory receptors is transformed orthogonally into horizontally interacting cortical fiber-mesh areas:

> I propose that interactions among the patterns of excitation which fall on receptor surfaces become, after transmission over pathways organized in a parallel fashion, encoded by virtue of horizontally interacting processes . . . . The hypothesis is based on the premise that neural representations of input are not photographic but are composed not only by an initial set of feature filters but by a special class of transformations which have considerable formal resemblance to an optical image reconstruction process devised by mathematicians and engineers. This optical process, called holography, uses interference patterns. It has many fascinating properties, among which the facility for distributing and storing large amounts of information is paramount.[96]

In contrast to the prevalent neurosurgical research paradigm that focused upon tracing the activity and wiring diagrams of single neurons to tease out the answer to the memory problem in the brain, Pribram began to search for evidence of some field or wavelike propagation in the junction of multiple neurons in the fine-fibered synaptic junction regions where complex dynamic electromagnetic fields could provide the vehicle for information patterning, processing, and storage.[97]

> The arrival of impulses at synaptic junctions is never a solitary event. Axonal terminations are usually multiple—i.e., axons branch at their ends. As many as 1,000 synapses may characterize the junctional possibilities between a pair of neurons. Dendrites are tree-like almost by definition, displaying many fine fibered branches which crisscross, making multiple contacts among neurons to form a pattern . . . . Inferences about the nature of such a pattern can be made from the known fine structure of the brain and the electrical activity recorded from it. Several such inferences suggest that these patterns make up wave fronts . . . an advancing wave may sweep over 100,000 neurons in a single second. Such a wave has rich potentiality, and the hypothesis that the mechanism of the junctional microstructure of slow potentials provides this organization.[98]

Pribram was also encouraged in 1964 when results of several independent research projects proved that the general consensus, which had asserted that there must be a single neuronal location for a memory engram, was determined to have been false.[99] Computer systems had detected three-dimensional patterns over clusters of cerebral neurons, revealing the possibility of distributed, parallel processing, rather than simply the previously assumed linear neuron-to-neuron processing. Pribram immediately related these patterns to the distributed geometric patterns he had observed in holograms, and he began referring to them as "holoscape contours."[100]

Pribram devised an experiment in which, through use of small spinning, notched gears, he measured the electrical response of a rat's buccal (lip) neuron to mechanical frequency stimulation of the rat's whiskers. The outcome revealed specific yet distinct complex three-dimensional contour patterns in response to different frequencies of whisker vibration.[101]

Such results strengthened Pribram's suspicion that perceptual processing would be found in patterns described by groups of neurons processing information in a spectral, frequency domain, rather than neuron-to-neuron impulse transmission:

> In my experience, each cortical neuron encodes several features of our visual environment, each neuron is like a person with many attributes, and thus our ability to recognize

> features had to depend on patterns displayed by groups of neurons for further processing.[102]

Pribram believed that he had found evidence that vision was undergoing two-way mathematical Fourier transforms in the processing of sight signals from photons striking eye cells to images perceived inside the brain. He relates a story of a conversation he had at the time, while climbing with colleagues on a hike in Colorado just prior to attending a neuroscience conference in Boulder:

> We had climbed high into the Rocky Mountains. Coming to rest on a desolate crag, a long meditative silence was suddenly broken by a query from Campbell: "Karl, do you really believe it's a Fourier?" I hesitated, and then replied, "No Fergus, that would be too easy, don't you agree?" Campbell sat silently awhile, then said, "You are right, it's probably not that easy. So what are you going to say tomorrow down there?" I replied, this time without hesitation, "That the transform is a Fourier of course!" Campbell smiled, chortling "Good for you! So am I."[103]

Yet even in light of published, repeatable evidence, this Fourier transform based theory met with substantial resistance within the mainstream research community. Neurophysiologists were reluctant to consider this mathematical, frequency field approach to perception and cognition, as they were already invested deeply in biochemical and physiological approaches. This led to what Pribram here describes, rather amusingly, as the formation of two camps among researchers, those captivated by "feature creatures" vs. the "frequency freaks":

> Within the research community the contrast between these two views of perception was wittily joked about as pitting "feature creatures" against "frequency freaks." Feature creatures were those who focused on "detectors" within our sensory systems, of features in our external environment—a purely "bottom-up" procedure that processes particulars into wholes. Frequency freaks, by contrast, focused on similarities between processes in the visual mode and those in the auditory (and incidentally the somatosensory) mode. For the frequency freaks, the central idea was that our visual system responds to the spectral, frequency dimensions of stimulation very much as the auditory system responds to the oscillations

> that determine the pitch and duration of sound. The approach of the frequency freaks, therefore, is "top-down"—wholes to particulars. Within the neuroscience community, the feature creatures still hold the fortress of established opinion, but frequency freaks are gaining ascendance with every passing year.[104]

In 1988, at age 70, Pribram retired from his position at Stanford as professor emeritus of psychophysiology, and relocated to Virginia to head the new Center for Brain Research and Informational Sciences (BRAINS) at Radford University. There he continued to do research and to give lectures in the areas of cognitive neuroscience and computational neuroscience.[105]

In the 1990s his theories continued to mature with the discovery of the Gabor function, in Paris, over a dinner with Gabor. What intrigued Pribram was not simply that Gabor found the Fourier transform so important in the mathematics of communication, but that he had pioneered an entirely novel new way of encoding and transmitting information, using what eventually came to be widely known as the "Gabor elementary function."[106]

When he discovered the Fourier holographic procedure, Gabor had been working on the problem of maximizing telecommunication across the newly laid trans-Atlantic cable.[107] He wanted to establish the *maximum* amount a message could be compressed without losing its intelligibility. To do this, Gabor developed the mathematics to model the two-way translations (movements) of Fourier flux (packets of information) through both space and time (in Gabor's application, a proposed undersea cable).[108]

This may be understood as the movement of the results of a Fourier transform, the transportation of a "packet" of frequency information, which normally has no time component, through time and space. Gabor designed a procedure to maximize information flow across the trans-Atlantic cable by first performing a Fourier transform on incoming frequency spectrum packets (typically milliseconds of audio conversation or music), and then transmitting the Fourier transformed information as electronic snapshot packets across the trans-Atlantic cable, where they could be received and de-encoded through application of a reverse Fourier transform.

A true mathematical Fourier transform requires an infinite range of frequencies, and in 1946 it was Gabor's challenge to determine the minimum number of frequencies that could be included in the transform to maintain intelligibility, while maximizing the amount of data being sent simultaneously across the cable.[109]

Over dinner, Gabor explained to Pribram how he solved the problem, using music as an example. While the pitch of a tone is dependent on the frequency of vibration, and there may be many harmonics accompanying that tone in the frequency spectrum, the duration of the tone complex is not measured in any of those frequencies but is instead measured on a completely different time axis. Accordingly, the function Gabor was seeking must include coordinates describing both a "windowed-packet" of tone information and *also* a set of space–time coordinates to map the packet flow.[110]

Gabor solved his problem through application of a "Hilbert space," a vector mathematical method capable of mapping an arbitrary number of dimensions, beyond the usual four dimensions of space (x_d, y_d, z_d) and time (t). The Hilbert space is named after the German mathematician David Hilbert (1862–1943), who in 1926 described a way to extend the methods of algebra and calculus to any number of dimensions. Since 1926, quantum theorists had been able to use Hilbert's approach to represent classical concepts within the new quantum framework, and use of the Hilbert space also offered a computable framework for the Fourier expansions required by quantum mechanics.[111] Hilbert, a close friend and associate of Albert Einstein, was reputed to have an acerbic sense of humor, and once, upon hearing that one of his students had dropped out of mathematics to study poetry, Hilbert was heard to reply, "Good, he did not have enough imagination to become a mathematician."[112]

The concept of a Hilbert space is essentially the consideration of a "space" having more than three dimensions, allowing the exploration of mathematical relationships involving an infinite number of dimensions (the subject is sometimes referred to as "infinite dimensional function spaces").[113] A simplified example of a Hilbert space is shown in Fig. 4 where the superposition of the two orthogonal dimensions found in the top half of the figure result in a Hilbert space of six dimensions shown at the bottom center.

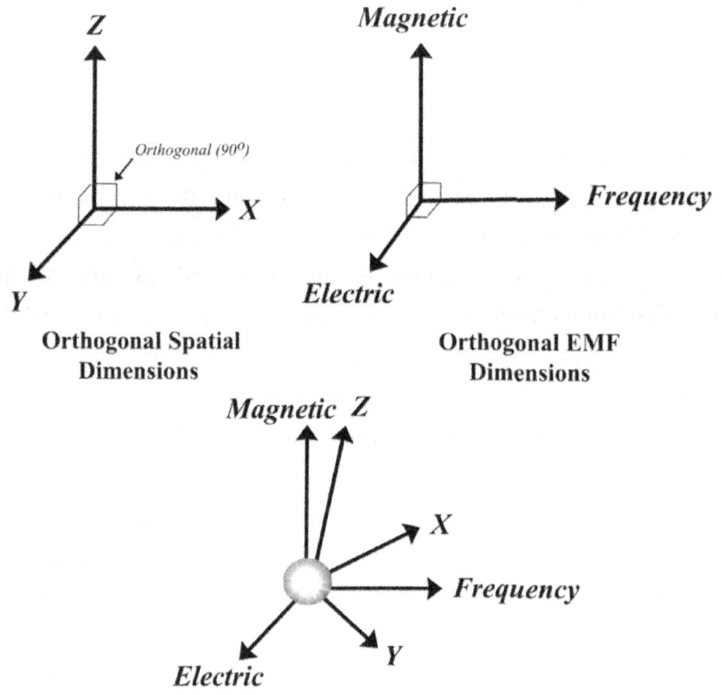

Figure 4. *A Hilbert space of six orthogonal dimensions.*

In the figure above, the "Hilbert space" is depicted at the bottom center, a "space" where all six dimensions now intersect at a common point, the "origin." Understanding this superposition became key for Pribram in expressing his holonomic theory, which he refers to as "deep processing," and states that the "the holonomic theory uses vectors in Hilbert phase space to express covariance."[114]

Although six dimensions do appear around the same origin in the Hilbert space in Fig. 4, (X, Y, Z, Electric, Magnetic, Frequency), the dimension of *time* is missing. In 1946 Gabor published a new mathematical approach which solved the problem of applying the

Fourier transform to a Hilbert space, by translating information from Hilbert dimensions into slices of time-based data. Gabor originally called his method the "windowed Fourier transform," but soon after publication it became known as the "Gabor function." Gabor had successfully devised an ingenious way to encode multiple "packets" of electromagnetic data and transmit them through the trans-Atlantic cable, simultaneous in both directions, to maximize transference rates and frequency content.[115]

Pribram was delighted; from that Paris dinner with Gabor had emerged a feasible mechanism with which to model the two-way flow of memory imprinting and retrieval in the brain, Gabor's "windowed Fourier Transform." Pribram related the process to that of the operation of a barcode system:

> Based on descriptions of the maps of receptive fields of cortical cells, I have proposed that the deep structure of our memory, as it is encoded in fine fibers in the brain, is much like a windowed Fourier transformation. The surface structure of memory, our retrieval and updating procedure, resembles what goes on at a checkout counter. It is only a step to conceive of a name as a barcode trigger identifying a purchase.[116]

Holoflux and the Frequency Domain

In evaluating the evidence that he had collected from hundreds of experiments in over 40 years of neurosurgical research at Yale, Harvard, and Stanford, Pribram observed that in the primate sensory systems, both visual processing and audio processing displayed holographic-like contour patternings among neuronal dendrites in the cerebral cortex. "In fact," he speculated, "the receptor mechanisms of the ear, the skin and probably even the nose and tongue work in a similar fashion."[117] With this new perspective offered by the Fourier transform and the unexpected posibilities for data storage found in the process of holography, Pribram, though always supported by laboratory data, pushed the envelope of his understanding forward:

> For me, theory is data-based and, I have, whenever possible, obtained in my own laboratory, first-hand, the data critical to theory.[118]

In contrast with the lack of success others were having in their efforts to establish the mechanism and location of memory engrams, Pribram put forth a theory that the brain mechanics of vision might be seen as holographic processes, projections onto the cortical surface from a frequency domain through the mechanism of frequency superposition and electromagnetic wave interference.[119] Pribram called this "the holonomic brain theory," and postulated the importance of the *frequency domain*:

> Essentially, the theory reads that the brain at one stage of processing performs its analyses in the *frequency domain* . . . a solid body of evidence has accumulated that the auditory, somatosensory, motor, and visual systems of the brain do in fact process, at one or several stages, input from the senses in the *frequency domain*.[120]

Pribram's theory holds that the mathematics of the Fourier transform operates within the brain, modulating electromagnetic fields of flux to create, store, replay, and process holograms within some as yet unidentified physiological context or substrate, but one that Pribram conjectures may be found within the fine-fibered dendritic webs of cerebral cortex regions.[121]

Pribram clearly distinguishes between observations made in the frequency domain and those made in space and time. While the ontological realization of flux in space–time can be conceptualized in the gyrations of three-dimensional electromagnetic waves, Pribram defines the ontological substrate of spectra in the frequency domain to be pure spectra, or *holoflux*, and he states that holograms are prime examples of *holofux*:

> The spectral domain is characteristically a *flux*, composed of oscillations, fluctuations, whereas interference patterns among waves intersect to reinforce or cancel. Holograms are examples of spectra, that is, of a *holoflux*.[122]

Thus the *holoflux* paradigm is essentially the same in concept as Bohm's *holomovement*, though in "Brain and Mathematics," Pribram takes exception to Bohm's use of the words "flow" or "movement," holding that such words cannot be used to characterize dynamism in a dimension devoid of space or time axes:

> David Bohm (1973) had a concept similar to flux in mind, which he called holomovement. He felt that my use of the term "flux" had connotations for him that he did not want to buy into. I, on the other hand, felt holomovement to be vague in the sense of asking, "what is moving?"[123]

Perhaps the flow issue can be resolved in the model of a Hilbert space; this can be accomplished by the superposition of the orthogonal dimensions of space (XYZ) and the electromagnetic frequency field (electromagnetic) in the movement of holoflux packets along the dimension of time (Fig. 5, "Flux packets moving through space–time").

Within the frequency domain itself, the holoflux can be seen as neither static nor dynamic in the spatial and temporal sense, because the energy in the flux domain is understood (by definition) to be "outside of time" and "outside of space."

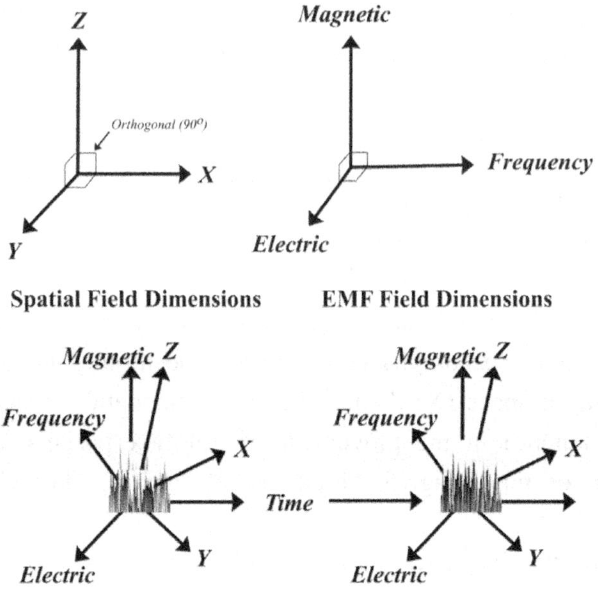

Holoflux Packets in A Hilbert Space

Figure 5. *Flux packets moving through space–time.*

These conditions of the frequency domain correlate with Bohm's concept of the "implicate order," the enfolded, sub–quantum dimension of his ontological quantum theory.[124] While Bohm's theory is examined in detail in Chapter 5, here a parallel concept can be visualized in Pribram's flow of holoflux energy from out from a frequency domain into actualized phenomena in space and time:

> Only through their manifestations in space and time do we get to know of the existence of potentials such as those of energy, of momentum, and of holographic processes. We can know the potential domain only by way of realization in space and time.[125]

Pribram regarded frequency as the direct mathematical link between a space–time electromagnetic domain and a transcendent flux domain, the mathematics of frequency bridging the two through a Fourier transform process. He goes so far as to say, "The frequency approach . . . is essentially Pythagorean."[126] At a San Francisco Zen

Center physics and consciousness conference Pribram came across a conceptual diagram of the Fourier transform process (Fig. 6), which perfectly encapsulated what he saw to be the function of the Fourier Transform, linking a spectral (frequency) domain to a space–time domain via the mathematics of the Fourier transform.[127]

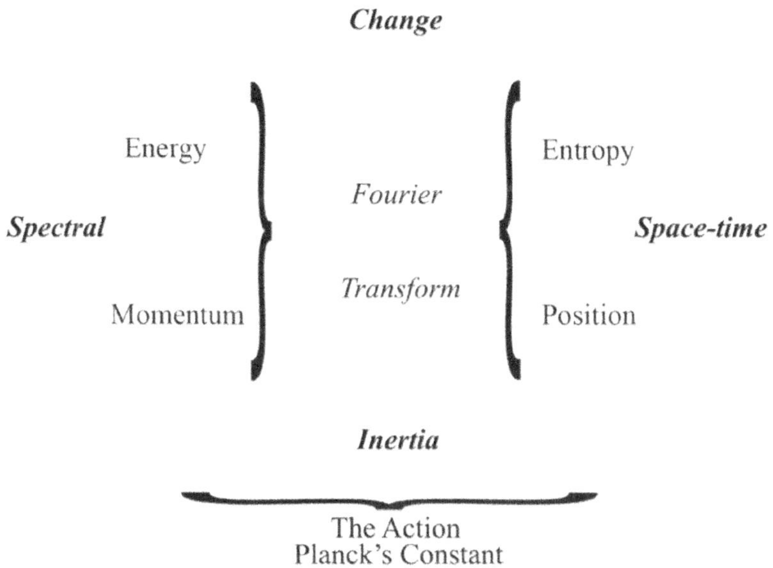

Figure 6. *Pribram's Fourier Transform diagram.*

Pribram first encountered this diagram during a presentation given by the physicist Geoffrey F. Chew, a Berkeley theoretical particle physicist, who told Pribram that he himself had received it from his colleague at Berkeley, Henry Stapp, who himself said he had been given it directly from Paul Dirac. Whatever the origin of the diagram, Pribram chose to publish it in several future papers. In his essay, "Consciousness Reassessed," Pribram's caption to the figure reads, "The Fourier transform as the mediator between spectral and space–time descriptions."[128] In "Brain and Mathematics," on the other hand, Pribram captions the figure, "The wave/particle dichotomy is orthogonal to the above distinction."

In Dirac's diagram, the spectral domain is positioned to the left and space–time to the right. The diagram can be seen as a key to

Pribram's model, and also posits that the Fourier transform process operates at a boundary dimension of Planck's constant, termed in the diagram, "The Action: Planck's Constant." While Pribram himself never discusses what "The Action: Planck's Constant" might imply for the diagram and his theory, the concept is perfectly congruent with David Bohm's discussion of the Planck length as boundary between the implicate domain and the explicate domain (discussed in Chapter 4). This Fourier diagram, according to Pribram, has clear implications:

> Matter can be seen as an "ex-formation," an externalized (extruded, palpable, compacted) form of flux. By contrast, thinking and its communication (minding) are the consequence of an internalized (neg-entropic) forming of flux, its "in-formation." My claim is that the basis function from which both matter and mind are "formed" is flux (measured as spectral density).[129]

An enhanced depiction of the spectral domain and the space–time domain is presented in Fig. 7 where electromagnetic fields and matter can be seen in the space–time domain to the right, and holoflux fields and frequency flux can be seen to the left, separated by the bi–directional Fourier transform.

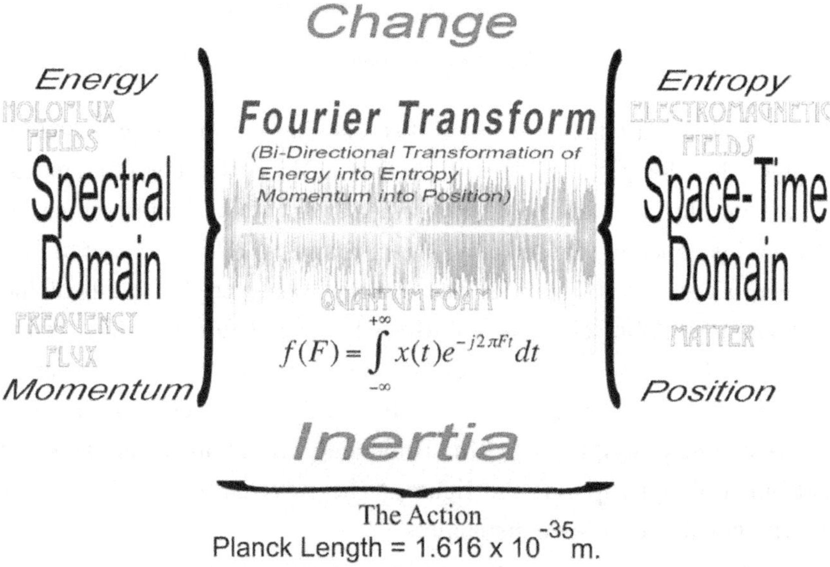

Figure 7. *Spectral domain and space–time domain.*

The notation at the bottom of the diagram suggests that this transformation occurs at the Planck Length limit of 1.616×10^{-35} m. Bohm here comments on holograms and the spectral domain:

> Spectra are composed of interference patterns where fluctuations intersect to reinforce or cancel. Holograms are examples of the spectral domain. I have called this pre-spacetime domain a potential reality because we navigate the actual experienced reality in spacetime.[130]

How then might we model the dynamics of the process behind the transformation (or collapse) of holoflux packets from the spectral domain (Bohm's enfolded cosmos) into the space–time brain structure? In a 1997 paper published in the *Proceedings of Symposia in Applied Mathematics*, Pribram suggests the following answer:

> An answer to the questions as to how mind becomes organized by brain rests on our understanding of the lessons of quantum mechanics and especially of that aspect which encodes the spectral domain, the implicate order. Although engineers daily use the spectral domain in radar, crystallography and tomography—wherever image processing is important—cognitive neuroscientists are, as yet, only barely acquainted with the pervasive nature of this order. It is now necessary to make accessible, both by experiment and by theory, the rules for "tuning in" on the implicate domain so that this domain can become more generally understood and scientifically validated.[131]

Another model of the spectral-space-time domains can be seen in Fig. 8, which suggest that the Gabor function, operating in parallel with the Fourier transform, describes mathematically the flow of holoflux packets moving back and forth simultaneously between the spectral frequency domain and the space–time domain.

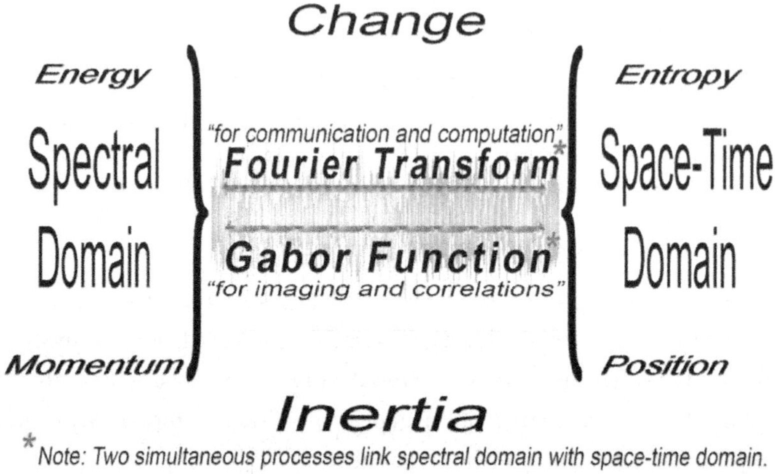

Figure 8. *Gabor and Fourier transform processes.*

An interesting theory, a proposal that provides a digital "clock" to describe the timing of this process of movement between space–time and the spectral domain, can be seen in the "theory of laminated space–time," proposed by Dewey in 1985.[132]

Dewey's theory proposes that energy vibrates digitally between two domains at a sub-quantum dimensional level, alternating between cyclical transformations, like the sequential ticks of a clock, between the spectral (frequency) domain and the space–time domain. The timing of such "ticks" as measured in the space–time domain, can be calculated, and corresponds to the Wien calculated frequency derived from this smallest possible wavelength, the Planck distance itself, or 1.616199×10^{-35} m. The frequency corresponding to such a wavelength would be at the limit of the frequency domain's extrusion into space–time, in other words, it must be the highest frequency (and correspondingly, the smallest wavelength) within the space–time domain. Dewey's theory visualizes the process of this flux extruding into space–time as a series of shells moving out at the speed of light, each shell a brane separated from the next by a spatial gap equal to the Planck length, as described here:

> I believe the universe to be composed of *nothing but* shells of electromagnetic particles which the theory of Laminated Spacetime describes as laminae of spacetime.[133]

In pioneering their holographic method, Emmet and Upatnieks used laser-based holographic technology to simplify the generation of a hologram by: (1) using a laser generated beam of photon energy (a *single frequency*), and (2) illuminating the object from two distinct fixed points in space (a *single, fixed phase-angle*). An interference pattern is created from the spherical waves of radiation incoming from the two fixed points in space, rippling at exactly the same frequency in space–time.

In nature, however, the situation is vastly richer in complexity: one can imagine radiant holoflux interactions processing in the space–time domain as a highly complex interaction of dynamic shells of every space–time radiation frequency conceivable; these shells impinge from an infinity of phase angles, like inwardly–directed radii aimed toward a central point. As contrasted with the human generated holograms of a single frequency of electromagnetic flux and a single vantage point (phase angle), each point in space–time actually experiences an *entire frequency flux spectrum* impinging on a particular point in space–time. In nature, all frequencies of the spectrum and all incoming angles of approach must be considered in the process configuration of the holoflux considered by Pribram and Bohm.

That such a process might involve quantum-scale phenomena as well as super–conductivity has been strongly suggested, not only by Pribram, but also by two of his colleagues:

> Mari Jibu and Kunio Yasue, an anesthesiologist and a physicist from Notre Dame Seishen University in Okayama, Japan, and I have suggested that the membranes of the fine fibers, the media in which the receptive fields are mapped, may have quantum-scale properties. The membranes are made up of phospholipids, oscillating molecules that are arranged in parallel, like rows of gently waving poplar trees. The lipid, fatty water-repellent part makes up the middle, the trunks of the "trees," and the phosphorus water-seeking parts make up the inside and outside of the membrane. These characteristics

suggest that water molecules become trapped and aligned in the phosphorus part of the membrane. Aligned water molecules can show super-liquidity and are thus super-conductive; that is, they act as if they had quantum-scale properties.[134]

In 1979, Karl Pribram, at that time a Stanford professor, was introduced to David Bohm, Fritjof Capra, and Brian Josephson at a conference in Cordoba, Spain, where the four presented papers.[135] At the time, Bohm was Professor of Theoretical Physics at London University, Capra, a theoretical physicist, was conducting research at the Lawrence Berkeley Laboratory, and Josephson a theoretical professor from Oxford who had been presented the Nobel Prize in 1973 for his work on quantum tunneling and superconductivity (and who had become a serious student of transcendental meditation in 1971).[136]

During the conference, Pribram discovered that Bohm's model of the implicate order and its projection, or extrusion, into space–time corresponded surprisingly well with his own holonomic theories. This became evident, in particular, when Bohm described how the implicate unfolds into the explicate and then the explicate enfolds into the implicate in a continual two-way process:

> The implicate order can be thought of as a ground beyond time, a totality, out of which each moment is projected into the explicate order. For every moment that is projected out into the explicate there would be another movement in which that moment would be injected or "introjected" back into the implicate order. If you have a large number of repetitions of this process, you'll start to build up a fairly constant component to this series of projection and injection. That is, a fixed disposition would become established. The point is that, via this process, past forms would tend to be repeated or replicated in the present, and that is very similar to what Sheldrake calls a morphogenetic field and morphic resonance. Moreover, such a field would not be located anywhere. When it projects back into the totality (the implicate order), since no space and time are relevant there, all things of a similar nature might get connected together or resonate in totality. When the explicate order enfolds into the implicate order, which does not have any space, all places and all times are, we might say,

> merged, so that what happens in one place will interpenetrate what happens in another place.[137]

In 1984 Pribram and a colleague obtained laboratory proof that cells respond to specific bandwidths of frequencies, further support for his holonomic theories:

> Together with an engineering student, Amand Sharafat (Pribram et al., 1984),[138] I performed an experiment in which we investigated whether neurons in the cat motor cortex were tuned to certain bandwidths of frequencies of passive movements of their forelimbs. Here, for once, we were testing a specific hypothesis, and the hypothesis was supported by our results. Certain cells in the motor cortex are responsive to the frequency.[139]

In a 1997 paper which he presented at a symposium for applied mathematics, Pribram offers a view even more congruent with Bohm's theories, hypothesizing that the mind–brain ontology is monistic, that a duality between mind and matter only *appears* to manifest because information that can be measured is unfolded into observable space–time coordinates from an enfolded potential.[140]

> According to this view, another class of orders lies behind the level of organization we ordinarily perceive. The ordinary order of appearances can be described in space–time coordinates. The other class of orders is constituted of fine-grain distributed organizations, which can be described as potential in the Aristotelian sense because only after "radical" transformation is their palpability in spatiotemporal terms realized. When the potential is actualized, information (the form within) becomes unfolded into its ordinary space–time manifestation; in the other direction, the transformation enfolds and distributes the information much as this is done by the holographic process . . . . Thus, on the one hand, there are enfolded potential orders; on the other, there are unfolded orders manifested in space–time.[141]

But the model of mind/brain as a quantum mechanical process poses a new question. In 2002 Pribram presented a paper in Torino, Italy, in which he discusses the application of the Fourier basis of the holographic principle as a useful metaphor in describing the brain/mind relation with regard to perception.[142] In this paper he poses the following question:

> Does this application indicate that the formalism of quantum physics applies more generally to other scales of inquiry? Alternatively, for brain function, at what scale do actual quantum physical processing take place? At what anatomical scale(s) do we find quantum coherence and at what scale does decoherence occur?[143]

Pribram himself discounted the possibility that electrical brain waves recorded by electrodes placed on the scalp might be significant carriers of cognitive information:

> There is some considerable doubt whether "brain waves" as presently recorded form the substrate of any meaningful interference pattern organization for information processing, although they may indicate that such process is taking place. The wavelengths recorded are, of course, considerably longer than those of light waves and can therefore only be carriers of small amounts of information—though in the form of spatially interfering holographic patterns.[144]

While Pribram never proposed an answer to his question as to what scale the holonomic process occurs in human physiology, the author proposes, in her own dissertation, that the anatomical scale required for Fourier holography lies in the infrared wavelength region.[145] Perhaps not coincidentally, the temperature spectrum within which the human mammalian body operates corresponds to a wavelength that perfectly fits the average inner diameter of blood capillaries, which, ranging from 8 to 10 microns, are found not only throughout the brain, but throughout the entire mammalian physiology.[146] Infrared waves in space–time have a wavelength of approximately 10 microns, but the spectral frequency associated with these waves, according to Pribram, lie in a non space–time spectral domain:

> Waves occur in space–time. Spectra do not. Spectra result from the interference among waves—and, by Fourier's insightful technique, the locus of interference can be specified as a number that indicates the "height" of the wave at that point. The number is represented by a Gabor function (or

other wavelet.) Frequency in space–time has been converted to spectral density in the transform domain.[147]

Reality of the Frequency Domain

But is this frequency domain "real" or is it simply a mathematical abstraction? This question has been at the heart of quantum physics since Erwin Schrödinger, often referred to as "the father of quantum mechanics," first described the wavefunction (ψ) in January 1926.[148] Prior to the discovery of the wavefunction, physicists were trying to understand the strange situation of experimental data that returned both wave-like and particle-like behavior.

> Schrödinger was happy to dispense with particles altogether. He viewed the wavefunction as the very real manifestation of a completely undulatory material world. In his description, particle-like behaviour was an illusion created by the overlapping and reinforcing of collections of "matter waves."[149]

However, the mathematician turned physicist Max Born (1882–1970), was soon to radically re-interpret the success of Schrödinger's wavefunction. In July 1926, just six months after Schrödinger's discovery, Born wrote a paper stating that:

> the wavefunction represents the *probabilities* that the electron wave will be scattered in certain directions . . . . A more precise consideration shows that the probability is proportional to the square of the wavefunction.[150]

Though Born's mathematically statistical calculations worked in quantum mechanical experimental predictions at the time, Einstein himself was not very happy with this probabilistic, statistical approach, and in a letter to Born later that year (December 3, 1926), Einstein remarked:

> Quantum mechanics is certainly imposing. But an inner voice tells me that it is not yet the real thing. The theory says a lot, but does not really bring us any closer to the secret of the "old one." I, at any rate, am convinced that *He* is not playing at dice.[151]

Much later in the century, Bohm, a young colleague and neighbor of Einstein's at Princeton, also adopted the realist view that the wavefunction directly describes a reality, not a set of probabilities. In developing his own quantum potential wavefunction (*Q*) in the 1980s, Bohm's primary assumption was that the wavefunction must be assumed to be real:

> The wavefunction is assumed to represent an objectively real field and not just a mathematical symbol . . . we suppose that there is, beside the field, a particle represented mathematically by a set of coordinates. We suppose that the particle is acted on not only by the classical potential but also by an additional "quantum potential."[152]

The debate between the two approaches to the wavefunction has continued into the 21st century, with the realist side supporting what has come to be called the *ψ–ontic* model (holding that it is the "real state of affairs") vs. the *ψ–epistemic* model group, which holds that the wavefunction should be regarded as a mathematical device only, a probability distribution).[153]

The debate has recently taken a turn in favor of the Bohmian realist approach; in early 2015, experimental proof supporting the ontological reality of the wavefunction was published in *Nature*, indicating that the quantum wave function may, indeed, be ontologically "real."[154] Experiments carried out at the University of Queensland consisted of repeatable measurements of a nonlocal wave function collapsing into a particle in space–time, yet the results failed to be predicted by probability distribution calculations. Single photons were manipulated in both horizontal and vertical polarizations, and their paths were controlled by an interferometer, while, according to the team, "every optical element in the experimental setup was carefully calibrated and characterized."[155] The paper, published in the January edition of *Nature,* is summarized as follows:

> Our results conclusively rule out the most compelling *ψ–epistemic* models . . . . This suggests that, if we want to hold on to objective reality, we should adopt the *ψ–ontic* viewpoint—which assigns objective reality to the wavefunction, but also has some intriguing implications such

> as non–locality or many worlds . . . . The only alternative is to adopt more unorthodox concepts such as backwards-in-time causation, or to completely abandon any notion of objective reality.[156]

The Brain as a Mere Computing Machine

Modern brain research in general has consistently approached study of the brain as if it were a computing machine operating only in space–time domains. In some sense this approach exhibits disinterest in actual consciousness or cognition per se. It is as if researchers are not focused upon consciousness but find it of greater urgency to explore how information might be stored, retrieved, and processed by the component parts of the brain acting as a computing machine.[157]

An example of this assumption can be seen in the Preface to *Neuroelectrodynamics: Understanding the Brain Language*, a textbook published in 2010. Here the authors on the one hand criticize the conventional neuronal spike pattern approach, and champion an electrodynamic field approach, yet on the other hand they continue to focus primarily on the paradigms of information processing of the brain as a computer:

> The brain is an incredibly powerful computing machine shaped by an evolutionary process that has developed over millions of years. The essence of brain function consists in how information is processed, transferred and stored. Understanding how our brain computes can explain why our abilities of seeing and moving in space are substantially more powerful as compared to any artificially built system . . . . However, from a neurophysiological point of view, current doctrine remains focused within a spike timing paradigm. This paradigm has limited capability for advancing the understanding of how the brain works . . . . Neuroelectrodynamics is a completely different model where the intrinsic computational processes are described by the dynamics and interaction of charges . . . . Spike timing including spike patterns can be regarded as an epiphenomenon or a result of these hidden processes.[158]

Pribram's Participatory Experiences

Among the more than 700 papers and books published by Pribram during his lifetime, there is almost no reference to any personal knowledge or opinion regarding religion, meditative exercises, or psychotropic drug experience.[159] However, late in his career he did discuss such topics, and from these comments we may be able to infer his own experience in these areas; such evidence can be found in interviews or panel discussion, but almost never in formal publication.

In a 1984 publication Pribram states that *consciousness* has now become central to the mind–brain problem due to "drugs, meditation, and a variety of other techniques."[160] It is certainly possible here that Pribram is alluding to his own first-hand participatory experience of such explorations, particularly as he sees them as "designed to promote psychological growth":

> Currently, the study of *consciousness* as central to the mind–brain problem has emerged from the explorations of altered and alternative states produced by drugs, meditation, and a variety of other techniques designed to promote psychological growth.[161]

An even stronger indication of first-hand participatory knowledge may be found in a remarkable paragraph, written at age 94, in which Pribram describes an experience of "holoflux." Here, his use of the phrases "we experience" and "we can experience" strongly implies direct knowledge of such states:

> Boundaries determine what is "without" and what is "within." This raises the issue as to whether the skin can really be considered a boundary between our self and the outside world . . . . Meditative exercises, drugs and dreams go further in dissolving these boundaries. First, we experience patterns rather than shapes and, if taken to the extreme, we can experience the loss of patterns as well: only a holoflux, "the white light" remains.[162]

There is also evidence of Pribram's association with the Transcendental Meditation movement. In 1977 a Stanford newspaper published an announcement of an upcoming presentations to be given by Pribram and his Stanford colleague, William Tiller, at an event sponsored by local Transcendental Meditation organizations.[163] One of the primary teachings of the Transcendental Meditation organization, under the guidance of Maharishi Mahesh Yogi, is the practice of mantra. It is possible to infer through this association that Pribram studied and practiced mantra himself.[164] Pribram specifically mentions the use of mantra in a 1991 publication, at the close of his book, in which he describes experiences of an enfolded, spectral dimension, in a somewhat cryptic final sentence:

> Perceptual experiences may on occasion however, reflect the spectral energy/momentum potential more than they reflect space–time configurations . . . *stimuli provided by a mantra*, for example. When the spectral dimension dominates the production of a perception, space and time become enfolded in the experienced episode. Time evolution ceases and spatial boundaries disappear . . . . The boundary between mind and matter, as all other boundaries, becomes dissolved. More on this at a future occasion.[165]

It is easy to argue that Pribram is talking about his own participatory experience of the contemplative practice of mantra, when he states "The boundary between mind and matter, as all other boundaries, becomes dissolved."[166] Pribram explains the experience of mantra stimulation in words that, as we shall see in Chapter 3, are virtually identical to Bohm's description of quantum dynamics ("space and time become enfolded"). While Pribram describes "perceptual experiences" as moving between "space–time configurations" and "spectral energy," he fails to say anything beyond this about his own perceptual experiences or altered states of consciousness. Nor does he follow up on his promise of "More on this at a future occasion." The rest of the book consists of a 55-page technical appendix written by Kunio Yasue and Mari Jibu on Quantum Brain Dynamics, mathematically supporting Pribram's conjectures on holonomic brain theory.[167]

Spirits as Small-Fibered Electrical Fields

More than 20 years after describing his experience of mantra in *Brain and Perception*, Pribram, in *The Form Within*, discusses his understanding of "spirit" in the context of his life's work:

> Spirit is ordinarily defined as being immaterial. And thermodynamics is a scientific endeavor that is based on the utilization of forms of energy, not matter. Also, descriptions of holoflux, (the quantum potential, zero-point energy) and quantum holography are devoid of any space–time structure. These endeavors are scientific and the patterns they investigate describe the patterns ordinarily probed under the rubric "spirit." There was a time when brain processes were described as mediated by "breath"— that is, spirit. Now we affirm that these "spirits" are electrical, and in the small-fibered web, they form electrical fields. And as we continue to explore the effects of magnetism and "soft photons," that is of heat, there is likely a great deal more we will discover about brain processing that affirms our spiritual nature. We know that Homo sapiens has experienced the spiritual aspect of "being" since the dawn of our species. Science is a quest and, as I have tried to show, there is no reason scientists should continue to restrict that quest to be confined merely to the composition of matter.[168]

But a more complete example of Pribram's interest in both Yoga and Zen, specifically in regard to his understanding of neurophysiology, can be found in a panel discussion held during the Science and Consciousness conference in Cordoba, Spain (1979), at which Pribram gave a presentation.[169] Table 1 shows that Pribram and Bohm presented at the conference, along with several other important thinkers in the field of physics and consciousness. It is possible that the Conference provided Pribram and Bohm their first chance to meet and exchange ideas, as Bohm was living near London at the time, and Pribram in California.

Table 1 - *Cordoba Conference Panel, 1979*

Participant	Title of Presentation
Fritjof Capra	"The Tao of Physics"
Karl H. Pribram	"Mind, Brain, and Consciousness"
David Bohm	"Imagination and the Implicate Order"
Brian Josephson	"Conscious Experience and Its Place in Physics"

During a panel discussion held at the Conference, Pribram provides a fascinating description of the difference between Zen and Yoga as mirrored in his own experience seen through the lens of his research in brain neurophysiology:

Brian Josephson: I would like to make a few comments on this question, especially with regard to the results of meditation . . . . Meditation can help those who practice it to find improved states of being. That said, meditation is not merely a technique, it also of necessity implies a spiritual dimensions and I do not think that we can ever measure spiritual values.

Karl Pribram: You are right on that point. The experiences of meditation are definitely of a spiritual nature, there is no doubt about that.

Actually there are, roughly speaking, two types of procedure used in the East, Zen and Yoga, and we are going to see how they show up in neurophysiology . . . . What we have done in our laboratory is to study the observation of recuperative cycles in primary sensory systems. In carrying out this research, we have realized that it was in fact possible to manipulate the recuperative time of the system by stimulating either the frontal or posterior part of the brain.

When the frontal part is stimulated, a much more rapid recovery is obtained—the channel acting as a synchronizer—and everything operates simultaneously.

> If, on the other hand, we stimulate the posterior part, we reduce the speed of the recovery of the system, but the channel is then of multiple type and carries much more information.
>
> I wonder, therefore, whether it is not this *difference* between the frontal and posterior parts of the brain that is operating in the various methods of meditation. That is, various techniques designed to induce modified states of consciousness, Zen appealing to the frontal part, and Yoga to the posterior.[170]

In an interview for *Psychology Today*, Pribram expands on his belief that, approached with the understanding that a holographic system of consciousness is operating in the universe, certain otherwise-unexplainable phenomena such as synchronicity, nonlocality, and paranormal phenomena can be explained:

> It isn't that the world of appearances is wrong; it isn't that there *aren't* objects out there, at one level of reality. It's that if you penetrate through and look at the universe with a holographic system, you arrive at a different view, a different reality. And that other reality can explain things that have hitherto remained inexplicable scientifically: paranormal phenomena and synchronicities, the apparently meaningful coincidence of events.[171]

In his autobiography published in 1998, Pribram discusses his affirmation that there lie other "classes of order" behind the conceptually-ordering map of our world given to us by Euclidean Cartesian space–time coordinates and Newton's laws; here can be seen, certainly in the phrases "enfolded potential orders" and "unfolded orders manifested in space–time," the influence and adoption of the ontological ideas of his friend and colleague, the quantum theorist David Bohm:

> This other class of orders constitutes distributed organizations described as potential because of their impalpability until radical changes in appearance are realized in the transformational process. When a potential is realized, information (the form within) becomes unfolded into its ordinary space–time appearance; in the other direction, the transformation enfolds and distributes the information, as this is done by the holographic process. Because work is involved

> in transforming, descriptions in terms of energy are suitable, and as the form of information is what is transformed, descriptions in terms of entropy (and negentropy) are also suitable. Thus, on the one hand, there are enfolded potential orders; on the other, there are unfolded orders manifested in space–time.[172]

And on the closing page of his final work, published in his 92nd year, Pribram ends with the following affirmation on the future scientific investigation of consciousness as it manifests in the domains of what is commonly called spirituality:

> Much of the research that I have reported here has dealt with forms, with patterns, that are, in themselves, not matter per se. These patterns can be considered the spiritual complements of matter. In this sense, spirit and spirituality now have status as topics ripe for scientific investigation.[173]

Pribram developed extensive laboratory data to support his paradigm that neuronal systems in the cerebral cortex and in the sensory systems of vision, hearing, and smell, operate on holographic principles. His contention is that such activity is not primarily in and among the neurons themselves, but that the process manifests within an extended electromagnetic field among the fine-meshed dendritic fibers of the cerebral cortex. He found evidence to support his contention in the outer reaches of the cerebral hemisphere, where he mapped detectable holographic transformations which led him to the conclusion that there exists an on-going "language" of transformation between space–time information and frequency-based information in the brain.[174]

In view of the serious nature of his discoveries during a lifelong search to understand the dynamic of memory and consciousness in the brain, it is refreshing to read Pribram's published answer to his own question of "what all this research is good for," expressed here at the age of 94:

> There are more than 37,000 neuroscientists currently performing experiments that address the topics that have formed my scientific life and the chapters in this book. From them we can expect a torrent of research results that bear on the health of our bodies and our brains. But there is more we

> need to know. I am often asked what all this research is good for? My answer is the one Faraday gave when asked this question regarding his discoveries of the properties of electricity: "Someday you'll be able to tax it."[175]

While Pribram and his associates in California were trying to understand a holonomic Fourier transform paradigm of brain processing (from direct experimental evidence and first-hand observation), a complementary paradigm was emerging in London, in the work of a theoretical physicist David Bohm, who began using concepts from quantum field theory and the Fourier transform to develop and support the cosmology of what he called the "implicate order."[176]

Chapter 3: David Bohm

Strong support for Pribram's holonomic theories relating brain and consciousness can be found in the quantum physics of the American physicist David Joseph Bohm (1917–1992), a Professor Emeritus of Theoretical Physics at Birkbeck College, University of London. Bohm's early research career began at the University of California in Berkeley, where he completed his dissertation in plasma physics in 1944 under Robert Oppenheimer at the Lawrence Radiation Laboratory in Berkeley.[177]

Much as Pribram's work was a search for a mind/brain process that was missing in mainstream research approaches, Bohm's search was a focus on something that was missing in quantum theory, the relationship between matter and consciousness. Bohm derided physicists who seemed to show no interest in the *meaning* of the quantum world, those who used quantum mechanics merely as a mathematical tool for effectively predicting results in nuclear physics experiments. At the same time, Bohm perceived that a fuller reality lies beyond appearances; that it is the creative process of the "whole," beyond theorizing, that provides the repeatable appearances of scientific measurement:

> The quantum jump or quantum leap is a creative process . . . . I called it the implicate order because it's the order of enfoldment that counts not the movement in a line. It manifests in the explicate. What we thought was the essence is now the appearance . . . . Namely, that the particle which was thought to explain the reality is now seen as an appearance. The essence of the true being is unknown. Even the implicate order is merely a concept, so even that should turn out to be an appearance. Though we agree that by examining these deeper, ever more penetrating appearances we can understand better, we are never going to grasp the whole. Even a unified theory will only be an appearance.[178]

In his approach, Bohm stresses that the distinction between information and meaning is an important one, as did Pribram (particularly in his *Languages of the Brain*, 1971). Unlike many physicists, whom he criticized for being information intoxicated, Bohm felt it of the utmost importance to search for meaning behind the information, and not to regard quantum mechanics simply as some exotic branch of calculation methodology. Understanding the underlying ontology behind quantum mechanics, Bohm felt, was a key to understanding the relationship between matter and consciousness. Here Bohm criticizes Bohr, the founder of quantum mechanics, who had won the Nobel Prize in 1913 for his quantum theory in modeling the hydrogen atom:

> Bohr's approach is to say nothing can be said about quantum mechanics at all other than to use it is a calculator, but my approach is to give it another appearance, more meaning. In the mechanical order it is hard to get much meaning.[179]

In the middle of his career as a theoretical physicist, Bohm's search for the underlying meanings to quantum theory was enriched through an introduction to a contemplative approach to the mind and consciousness expressed in the ideas of Jiddu Krishnamurti. In 1961, while a professor of physics at Bristol University, Bohm attended a conference in London to hear a series of Krishnamurti lectures on perception, a topic that, as a quantum physicist, greatly interested Bohm. During the prior year, Bohm had become interested in theories of consciousness and perception expressed in the writings of philosophers and mystics, and had begun reading books on Yoga, Buddhism, and the work of the mathematician P. D. Ouspensky, a follower of the mystic G. I. Gurdjieff.[180] Public knowledge of his new interest resulted in loss of credibility among Bohm's critics in the United States, as recounted here by his colleague and biographer, the physicist F. David Peat:

> When his former colleagues in the United States learned of this change of interest, it caused them considerable distress. In the years that followed, some lamented that Bohm had gone "off the rails," that a great mind had been sidetracked, and the work of an exceptional physicist was being lost to science.[181]

Bohm, however, remained completely serious in his search for an understanding of quantum theory, even though the quest had led him into these areas of philosophy, psychology, and mysticism. Bohm's driving interest can be seen in the first paragraph of *The Undivided Universe: An Ontological Interpretation of Quantum Theory*, completed in 1992, only weeks before he suffered a fatal heart attack; here he expresses his passion to understand the mysteries that continue to lie behind quantum theory, rather than regarding the theory simply as something "which we know how to use":

> The formalism of the quantum theory leads to results that agree with experiment with great accuracy and covers an extremely wide range of phenomena. As yet there are no experimental indications of any domain in which it might break down. Nevertheless, there still remain a number of basic questions concerning its fundamental significance which are obscure and confused. Thus for example one of the leading physicists of our time, M. Gell-Mann, has said "Quantum mechanics, that mysterious, confusing discipline, which none of us really understands but which we know how to use."[182]

To understand Bohm's approach to quantum theory it is useful to examine their development in the context of the complexity of his life. Born in Pennsylvania in 1917, Bohm took an early interest in science, having become an avid reader of science fiction by the age of seven. After graduating in physics and receiving the school's Mathematics Prize from the University of Pennsylvania, he accepted a fellowship to enter graduate studies at Cal Tech. Within a year he found himself dissatisfied with the seemingly exclusive focus on problem solving at Cal Tech, a focus which seemed to discount Bohm's passion for seeking an understanding of what must be underlying ontological realities, favoring instead the development of mathematical models and their manipulation.[183]

In 1941 Bohm met with J. Robert Oppenheimer at the University of California, Berkeley, having heard that in addition to being a brilliant theoretical physicist, the thirty-seven year old prodigy, Oppenheimer, encouraged wide ranging intellectual discussion among his grad students and postdoctoral researchers. Soon after this meeting, Bohm accepted a fellowship offer from Oppenheimer to study at Berkeley, and in 1941 Bohm moved from Los Angeles to

Berkeley, where he began graduate work in plasma research at the Lawrence Berkeley Laboratories.[184]

After a year of intense collaboration with other physicists at the laboratory, often working more than sixty hours a week, Bohm gave his first presentation of the ideas he had thus far developed to model the complex nature of plasma, but early in the presentation, as he began to try to convey his new ideas to Oppenheimer and the rest of his committee, something unexpected occurred:

> A naturally shy person, as he began to speak he nonetheless felt that everything was going extremely well. Soon he had the sensation that he was going beyond physics into something almost mystical, to the point where he felt himself in direct contact with everyone in the room. He was convinced that each individual consciousness had been transcended so that his audience was also sharing this experience. [185]

This intense feeling was overwhelming, leading immediately to a deep depression, and Bohm took a leave of absence, during which he was unable to work or study for almost an entire year.[186]

Even more problematical for Bohm during the first year of work on his dissertation under Oppenheimer was his growing interest in several activist antiwar groups, including the Committee for Peace Mobilization, the Campus Committee to Fight Conscription, and the Young Communist League, and "it was noted that Bohm had been involved in organizing activities for the Communist party and had distributed copies of Earl Browder's *Victory—and After*."[187] These associations and activities were later to haunt him.[188]

The Communist party in the United States had peaked at about 30,000 members in 1941, having grown steadily during the Great Depression due in part to the rapid growth of labor unions, but also because Communism was seen as a clear enemy of Fascism.[189] In November 1942, Bohm joined the Communist party in Berkeley, but membership was not particularly exciting or eventful, as Bohm later described to the historian Martin Sherwin during an interview: "The meetings were interminable, discussing all these boring attempts to stir up things on the campus which didn't amount to much."[190]

Back at Berkeley in 1943, Bohm completed his doctoral research project, an effort to develop a mathematical theory with which to model the dynamics of plasmas, a poorly understood phenomenon that greatly intrigued Bohm:

> As he studied the plasmas he became struck by their extraordinary nature. They began to take on, for him, the qualities of living beings. When physicists studied a plasma by introducing an electrical probe, it would generate a charged sheath around the probe and neutralize its effects. It was as if the plasma were protecting itself and preserving its internal status.[191]

Bohm completed his project, and the results were considered by Oppenheimer to be spectacular; the calculations that Bohm developed in his thesis were able to model accurately the behavior of uranium fluoride plasma in an electric arc, and led directly to an elegant technique for separating U^{235} from U^{238}, a critical step in the Oppenheimer's effort toward building a nuclear bomb.[192]

Unfortunately, shortly before the publication of his dissertation, Bohm's previous political activities caught up with him. It was suddenly apparent that his affiliation with Communism made it impossible for him to obtain a security clearance, while paradoxically the importance of his own dissertation for the war effort led to it being classified:

> The scattering calculations (of collisions of protons and deuterons) that he had completed proved useful to the Manhattan Project and were immediately classified. But without security clearance, Bohm was denied access to his own work; not only would he be barred from defending his thesis, he was not even allowed to publish his own thesis.[193]

Oppenheimer was accordingly forced to petition the University of California Board of Regents to rule on an exception to academic policy in order to approve Bohm's doctoral dissertation without allowing it to be published. The Regents assented and Bohm was awarded his doctorate and permitted to remain working at the Lawrence Laboratory, where he continued to focus his intense intellect upon the behavior of plasma flow and the mathematics of flux movement.[194] Here in an interview, Bohm describes his early interests and work with Oppenheimer during the period 1943–1946:

> When I worked at the Lawrence Laboratory, after taking my PhD, I became very interested in the electron plasma. This is a dense gas of electrons that exhibits radically different behavior from the other, normal states of matter and it was a key to much of the work the laboratory was doing at the time. My insights sprang from the perception that the plasma is a highly organized system, which behaves as a whole. Indeed, in some respects, it's almost like a living being. I was fascinated with the question of how such organized collective behavior could go along with the almost complete freedom of movement of individual electrons.[195]

Unfortunately, Bohm's early flirtation with Communism made the suppression of his dissertation the first of what would eventually be a series of challenges to his academic career. Nevertheless, in 1947 the young physicist was hired as an assistant professor by Princeton University. During the relocation to New Jersey, he found a room near campus, and soon met his neighbor, another Princeton professor, the 57-year-old Albert Einstein, who, like Bohm, objected to quantum mechanics for its "lack of an objective description."[196] Bohm's clarity on this topic was no doubt enhanced by the many discussions he had with Einstein during his tenure at Princeton.[197]

During his time at Princeton, Bohm also continued his research on the mathematics of plasma dynamics, while giving classes and public lectures on quantum mechanics, but he refused to accept the majority view which was, "and still is," in the words of his colleague, B. J. Hiley,

> that the precise nature of the conceptual changes is not important . . . . All that is needed is to work with the self-consistent mathematical formalism, which, in some mysterious way, correctly predicts the numerical results of actual experiments. After lecturing on the subject for three years, Bohm thought that this was not a satisfactory position to adopt so he decided to try to get a better understanding of the subject by writing a definitive textbook in which the physical aspects of the mathematics would be emphasized.[198]

Bohm worked furiously in his spare time on this major textbook which was to be a full interpretation of quantum theory, grounded both in mathematics and conceptual clarifications, in which he hoped would resolve major unanswered problems within the quantum theory.[199] He also began working on a reinterpretation of quantum theory in a paper of his own, which he called the "hidden variable theory." His textbook, *Quantum Theory*, was published in 1951, and his paper in 1952.

However, in October 1951, his reputation having been seriously impugned by the politics of McCarthyism, and under fear of being jailed for Contempt of Congress for failure to inform on his colleagues, Bohm fled the United States for Brazil.[200]

In late 1949, while working on the final chapters of his book, Bohm had been served with a subpoena ordering him to appear before Senator Joseph McCarthy at the House Un-American Activities Committee. At the hearing, Bohm was asked to identify which of his previous numerous colleagues at Berkeley had been interested in Communism. Not wishing to incriminate his friends and colleagues, Bohm repeatedly had his attorney plead his "right to silence" under the Fifth Amendment to the United States Constitution. Notwithstanding his constitutional rights, on December 4, 1950, Bohm was arrested at his home for Contempt of Congress, and though he made bail and was released, the administration at Princeton University peremptorily suspended him from "all teaching and other duties" for the duration of the trial.[201]

The negative publicity associated with the McCarthy hearings soon led to Bohm's dismissal from his position at Princeton, and subsequent, his inability to find employment anywhere in the United States. With letters of reference from Einstein and Oppenheimer, Bohm was offered, in 1951, a position in the physics department of the University of São Paulo, the largest university in Brazil. At the time the University was in the throes of development and expansion, during which had been seeking to attract a prominent quantum physicist from a mainstream American university.[202] Bohm decided quickly to accept the position and prepared to make a hasty departure for Brazil, a day described here by his biographer:

> On the October day in 1951 when Bohm left the United States, thunderstorms and hurricane-force winds swept across New Jersey and New York. The flooded streets made it difficult for him to reach the airport . . . . On board the aircraft while taxiing to takeoff, he heard an announcement that the plane would return to the terminal because of an irregularity in the passport of one of the passengers. Already highly nervous, Bohm wondered if he was going to be arrested. The problem, however, concerned another passenger, who was removed from the flight.[203]

In December 1951, two months after his arrival in Brazil, the US consulate in São Paulo confiscated Bohm's passport, effectively stripping him of his US citizenship, and making it impossible for him to leave Brazil other than to return to the United States.[204]

Undeterred, with his passion for learning undiminished, Bohm continued his work in São Paulo, and he was soon sufficiently fluent in self-taught Portuguese, sufficiently fluent to lecture at the University.[205] He was especially encouraged by favorable peer reviews of his recent US publication of *Quantum Theory*, the 628-page textbook interpreting quantum theory, and which begins in the first paragraph to emphasize the importance of familiarity with Fourier analysis:

> Modern quantum theory is unusual in two respects. First, it embodies a set of physical ideas that differ completely with much of our everyday experience, and also with most experiments in physics on a macroscopic scale. Second, the mathematical apparatus needed to apply this theory to even the simplest example is much less familiar . . . . It seems impossible, however, to develop quantum concepts extensively without *Fourier analysis*. It is, therefore, presupposed that the reader is moderately familiar with Fourier analysis.[206]

Theory of Hidden Variables

But perhaps the greatest significance of *Quantum Theory* lies in the fact that Bohm presents, this time in a major text book, his theory of hidden variables: "Perhaps there are hidden variables that really control the exact time and place of a transfer of a quantum, and we simply haven't found them yet."[207] Einstein's major objection to the quantum theory was its almost complete lack of an objective description, and it was this objectivity Bohm addressed in his new interpretation.[208]

At Princeton, Bohm and Einstein had agreed that there must be underlying processes yet to be discovered. Quantum mechanical calculations, while they work in the prediction of experimental evidence, treat particle actions collectively; in a similar way as taking temperature measurements are used to predict the motion of billions of gas molecules. In neither case can actual particle motion be mapped or predicted at the real dimensions being examined. Einstein and Bohm felt there must be principles that affect motion at a lower scale than that addressed by the collective mathematics of quantum mechanics. Bohm writes in his 1952 paper:

> This assumption has been the object of severe criticisms, notably on the part of Einstein, who has always believed that, even at the quantum level, there must exist precisely definable elements or dynamical variables determining (as in classical physics) the actual behavior of each individual system, and not merely its probable collective behavior . . . . The suggested interpretation provides a broader conceptual framework than the usual interpretation, because it makes possible a precise and continuous description of all process, even at the quantum level. This broader conceptual framework allows more general mathematical formulations of the theory than those allowed by the usual interpretation. Now, the usual mathematical formulation seems to lead to insoluble difficulties when it is extrapolated into the domain of distances of the order of 10^{-33} cm or less. It is therefore entirely possible that the interpretation suggested here may be needed for the resolution of these difficulties.[209]

In his paper on hidden variables can be seen Bohm's early fascination with the physics of the greatest depths of scale. Far below the normal range at which even quantum mechanics is applied, Bohm's fascination lay below, down in the scalar regions of the Planck length (approximately 10^{-33} cm), at a scale within which "classical notions such as causality or distance between events cannot be expected to be applicable."[210] The Planck constant is discussed in detail in Chapter 4.

Motion of "Particles" in Space–Time

Bohm explains apparent linear motion with an ontology derived from quantum theory when he speculates that, rather than manifesting as a solid, continuous particle in space–time, "the proton takes the form of a wave that collapses inward from all space," like a whirlpool in water.[211]

Similarly, in a conclusion that is contrary to the conventional assumption that electrons and protons are well-defined particles, Bohm describes the apparent motion of protons and electrons as:

> . . . a continuous process of inward collapse and outward expansion. Therefore, every elementary particle collapses inward from the whole of space. In fact, each elementary particle is a manifestation of the whole universe.[212]

Bohm's textbook on *Quantum Theory* was generally well received for its clarity of language, analogy, and supporting calculus, and was subsequently used in undergraduate courses on quantum mechanics worldwide. However the concept of *hidden variables*, described in the textbook and in his 1952 paper published exclusively on "hidden variables," was pounced upon almost universally by quantum physicists as being incompatible with the generally consensual prevailing "Copenhagen interpretation."[213] In Copenhagen, at a series of formal and informal meetings and conferences spanning the years 1925–1927, Niels Bohr and two of his young students, Werner Heisenberg and Wolfgang Pauli, had developed what was later to be called the Copenhagen Interpretation of quantum mechanics, a conjectural descriptive explanation of what the working mathematical equations seemed to imply.[214] To most physicists, having been in general agreement for over 20 years with the Copenhagen Interpretation, Bohm's hidden variable theory was seen as a direct challenge, almost a heretical slap in the face of the general consensus. However from Bohm's point of view, it was the Copenhagen approach, with its heavy reliance on mathematical probability theory, that was random and unsatisfactory.[215] In this, Einstein can be seen to be in full agreement, and he is quoted here in a 1926 letter to Max Born, where he comments on the recently published Copenhagen Interpretation, expressing special concern regarding the deep reliance of probability theory in quantum mechanics:

> Quantum mechanics is certainly imposing. But an inner voice tells me that it is not yet the real thing. The theory says a lot, but does not really bring us any closer to the secret of the "old one." I, at any rate, am convinced that *He* does not throw dice.[216]

The prevailing consensus among physicists held that quantum events are indeterminate and discontinuous, and therefore the physics of process is not linearly predictable at quantum scales of matter, and that only probability mathematics works in predicting outcomes. Bohm opposed this axiom with his conviction that the process of the universe cannot be attributed to purely random statistical activity, but that there is an underlying order and comprehensive connectivity, a unity to the cosmos, a wholeness not predicted by mere probability theory.[217]

In his hidden variable theory, however, Bohm assumes the universe to be coherent and predictable, and considers the apparent randomness to be misleading, proposing instead that there must exist hidden variables operating at sub-quantum levels, hidden forces and processes operating at lower levels below the currently assumed lower limit of matter, and not yet accessible with current instrumentation. Understandably, Bohm's new theory of hidden variables was immediately rejected by the community of physicists, in part for political reasons (Bohm's earlier communist affiliations), but primarily due to the firm attachment to the consensual acceptance of the Copenhagen Interpretation and the predictable results proven by use of quantum mechanical calculations.[218]

At best, Bohm's new theory was ignored; at worst, some went so far as to accuse him of being a Trotskyite and a traitor. One physicist termed Bohm's work "juvenile deviation," and another called Bohm "a public nuisance."[219] This strong reaction against Bohm's hidden variable hypothesis even led his own one-time supervisor, mentor, and hero, Robert Oppenheimer, to gather a meeting of physicists at Berkeley in 1953 with the It is as if researchers are not focused upon consciousness but find objective of disproving Bohm's theory; the meeting closed with Oppenheimer being heard to say, "If we cannot disprove Bohm, we must continue to ignore him."[220]

In 1955, Bohm left Brazil to accept a position as a physics professor at the Israel Institute of Technology in Haifa, where in 1956 he met (and married) his wife, an English artist by the name of Saral Woolfson. In 1957 Bohm and Saral left Israel where Bohm accepted a position at Bristol University in England. In London, that same year, Bohm presented his mature version of the hidden variable concepts, which he had first begun to develop in 1951 in *Quantum Theory*.

In his London paper Bohm introduced the idea of processes occurring at sub-quantum levels, operating at considerably smaller dimensions than those at which quantum mechanical calculations modeled the activity of atomic particles. This was the beginning of Bohm's conceptualization of the implicate order, through considering the possible ontological reality of dimensions far below the quantum dimensions being explored by quantum mechanics.

During his second year in England, Bohm continued to present a series of lectures at London University detailing his hidden variable theory; the lecture was attended by a young doctoral student in physics, John Bell, who had become interested in mathematical approaches to the nonlocality problem in quantum mechanics. Bell, encouraged by Bohm's theories, went on to develop a mathematical proof of nonlocality, a phenomenon that has since been verified several times in rigorous experiment.[221] It is of interest to note that Bell said, of Bohm's theory, "No one can understand this theory until he is willing to see psi as a real objective field rather than just a probability amplitude."[222]

Bohm did not limit his intellectual pursuit exclusively to physics. During his three years at Bristol University he began to read widely in areas of speculative philosophy and soon became acquainted with the process philosophy of Alfred North Whitehead, the Cambridge mathematician turned Harvard philosopher. One evening, according to Bohm, while reading *Process and Reality*, he experienced what seemed to be an epiphany; he had a sudden visualization of infinity represented in the image of a large number of silvered spherical mirrors, each one reflecting all of the others, a cosmos composed of an infinity of reflections, and of reflections of reflections: "Every atom was reflecting in this way, and the infinity of these reflections was reflected in each thing; each was an infinite reflection of the whole." [223] The influence of this image can be found 20 years later in Bohm's understanding of what he had by then come to call, "the Implicate Order." The image of mirrors reflecting the whole is echoed in *Wholeness and the Implicate Order* in the following passage, which describes Bohm's concept of the holomovement:

> In certain ways this notion is similar to Leibniz's idea of monads, each of which "mirrors" the whole in its own way, some in great detail and others rather vaguely. The difference is that Leibniz's monads had a permanent existence, whereas our basic elements are only moments and are thus not permanent.[224]

At Bristol in the early 1960s, during his wide-ranging reading in philosophy and Eastern metaphysics, Bohm was introduced to the writings of the Russian mathematician P. D. Ouspensky, who had written several books on the metaphysics of the mystic G. I. Gurdjieff.[225] In 1962 Bohm managed to meet in London with J. G. Bennet, himself a mathematician as well as an authority on Gurdjieff; the two corresponded over the next several years, but "their connection was broken," as Bohm's biographer says, in a footnote:

> Bohm became increasingly critical of Gurdjieff's emphasis upon the essential separation of an observing consciousness.

> For Bohm there was a "total awareness" that lies beyond consciousness and is the field in which creativity operates.[226]

Krishnamurti and Meditation

It was also while at Bristol University that Bohm first became acquainted with the philosopher Jiddu Krishnamurti. In the Bristol public library, Bohm's wife, Saral, discovered a book in the philosophy section discussing the topics of perception and thought, and brought it home as of possible interest. Bohm was immediately fascinated by what Krishnamurti had to say:

> My first acquaintance with Krishnamurti's work was in 1959 when I read his book *First and Last Freedom*. What particularly aroused my interest was deep insight into the question of the observer and the observed . . . I felt that it was urgent for me to talk with Krishnamurti directly and personally as soon as possible. And when I first met him on one of his visits to London, I was struck by the great ease of communication with him, which was made possible by the intense energy with which he listened and by the freedom from self-protective reservations and barriers with which he responded to what I had to say . . . it was in essence of the same quality as that which I had met in contacts with other scientists with whom there had been a very close meeting of minds. And here, I think especially of Einstein who showed a similar intensity and absence of barrier in a number of discussions that took place between him and me. After this, I began to meet Krishnamurti regularly and to discuss with him whenever he came to London.[227]

In 1961, Bohm left Bristol University to accept the offer of Chair of Theoretical Physics, offered to him by J. D. Bernal, the head of the physics department (and incidentally, a dedicated Marxist) at Birkbeck College, University of London. In that same year Bohm, for the first time, arranged a private conversation with Krishnamurti in London. The two continued their dialogues, at first privately, and then eventually in public venues, before audiences in London, Switzerland, and later in Ojai, California.[228]

In Switzerland in 1965, the first public recordings were made of what would soon become a series of dialogues between Krishnamurti and Bohm. Undoubtedly Bohm's previous reading experience of Whitehead, Hegel, Ouspensky, and Indian philosophy had prepared him well for these flowing dialogues with Krishnamurti, and likely influenced his own conceptualizations. In addition to published books containing transcripts of many of these dialogues, numerous video and audio recordings have been released; a recent YouTube search on "Krishnamurti" and "Bohm" pulls up 32,000 results.[229]

It was in 1967, during the "Summer of Love," that Bohm and his wife, at the urging of Krishnamurti, both became vegetarians and took up the daily practice of meditation. Bohm was particularly interested in Krishnamurti's teaching on the process of "dying to thought" in which the physical brain no longer gives energy to the movement of thought process in the brain.[230] According to Krishnamurti, it is this incessant conditioned mental thought process that masks and conceals the silence within which the less conditioned consciousness process might be perceived, and within which new modes of "non-verbal thought" might operate. Bohm here describes his own understanding of Krishnamurti's teaching on the process of meditation:

> Krishnamurti has observed that the very act of meditation will, in itself, bring order to the activity of thought without the intervention of will, choice, decision, or any other action of the "thinker." As such order comes, the noise and chaos which are the usual background of our consciousness die out, and the mind becomes generally silent . . . . In this silence, Krishnamurti says that something new and creative happens, something that cannot be conveyed in words, but that is of extraordinary significance for the whole of life. So he does not attempt to communicate this verbally, but rather, he asks of those who are interested that they explore the question of meditation directly for themselves.[231]

It is easy to imagine how Bohm's exceptional, trained capability for focus on abstractions (as a theoretical physicist) served to reinforce his growing capabilities in the practice of formless meditation. Bohm soon established a daily period in the early morning during which he would walk slowly in the park or countryside while watching the movement of this thoughts, not allowing himself to become distracted or "caught up" in their content It was a practice which he would throughout the rest of his life.[232]

The Implicate Order

It is of interest to note that within two years of beginning his daily meditation walk, Bohm conceptualized a significantly new idea, the "implicate order," which ontologically strengthened his hidden variables theory. By 1969 the subjects of topology, nonlocality, and pre-space had come to the forefront of discussion within the physics department at the University of London, and as his colleague, the physicist B. J. Hiley recalls, Bohm had a breakthrough in his understanding: "Suddenly, a new idea emerged from Bohm, apparently entirely out of the blue . . . Bohm called it the *implicate order*."[233]

Bohm identified this implicate order as being the source of the hidden, sub-quantum variables which he hypothesized were providing the causality underlying the operation of quantum mechanics, and he began an effort to model the effect of the implicate order mathematically.[234] In 1925, Werner Heisenberg had taken a similar view:

> Heisenberg, who at age twenty-four, was young even by physics standards, tried to save classical mechanics by abandoning it at Nature's bottom rung. Inside the atom, he declared, not only do particles and electron orbits have no meaning, but neither do even such basic classical properties as position, momentum, velocity, and space and time. And because our imaginations require a space–time container, this atomic world cannot be easily pictured.[235]

Heisenberg's approach, however, fell by the wayside when the Copenhagen approach proved so successful in predictive calculations of nuclear particle trajectories using probabilities. Decades later, Bohm pursued a similar idea, and in 1980 published *Wholeness and the Implicate Order*, in which is revealed both the mathematics and a mature conceptual model for his proposed implicate order, a model based not on "things" considered as objects, nor restricted solely to phenomena in a Cartesian space–time order. His new model understands reality as a lens-like flowing of simultaneous enfolding and unfolding dimensions between two orders, an implicate order and an explicate order, as Bohm explained:

> There is the germ of a new notion of order here. This order is not to be understood solely in terms of a regular arrangement of *objects* . . . . Rather, a *total order* is contained, in some *implicit* sense, in each region of space and time. Now, the word "implicit" is based on the verb "to implicate." This means "to fold inward." So we may be led to explore the notion that in some sense each region contains a total structure "enfolded" within it.[236]

Bohm went so far as to describe the implicate order as the ground within which the entire universe is enfolded at each and every "point" in space–time, as the Finnish philosopher Paavo Pylkkänen, a former student and friend of Bohm's, explains clearly:

> Just think of all the atoms and particles that constitute your body. We are used to thinking about them as tiny little things that just passively sit there. But quantum field theory, as interpreted by Bohm, suggests otherwise. There is a sense in which each particle in your body enfolds information about the whole universe . . . . There is also a sense in which information about each particle in your body is enfolded throughout the universe.[237]

However, Bohm's focus of interest clearly ranged beyond the conventional boundaries of physics, into theories of mind, thought, and consciousness. It is easy to conclude that Bohm's dialog with Krishnamurti influenced the expansion of this focus, bringing to the forefront for Bohm issues of thinking and mind. With his deep understanding of quantum theory, Bohm was able to see a direct connection between thought, consciousness, and the implicate order, as can be discerned in his description of the thinking process, recorded in a 1990 interview:

> And that's how the thinking process goes, it's enfolded in your consciousness, it unfolds to a certain thought, folds back, and then the next thought appears, different; a series of thoughts not too different seems to be continuous.[238]

In imagining this process as an "enfoldment," we are led to the obvious next question, "Enfoldment into what?" Bohm's non-Cartesian answer, then, is enfoldment *into something within*, or enfoldment *into* the implicate order. How then might this be conceptualized? Imagine the universe at the lowest possible spatial range. Moving to ever-smaller dimensions, to the very bottom of the dimensional scale, we encounter the end of space, according to accepted quantum theory, at a limiting length, below which length has no meaning (as discussed shortly).[239] Thus this implicate order, located at the center, everywhere, can be seen as the ground, base, or center out of which dynamically springs the space–time continuum itself, the explicate order.

The Sub-Quantum Order

Bohm applied the term "sub-quantum order" to that order of physics which he viewed as expressing the underlying causal reality *from which* quantum mechanics operates, calling this approach "going beyond the quantum theory."[240] While quantum mechanical theory works well for predicting the path of nuclear particles in dimensional ranges of around 10^{-15} meters, the theory places a heavy insistence on pure probability considerations. This randomness had always bothered Bohm, and as early as 1952, in his textbook, *Quantum Theory,* Bohm states:

> We have come to the point of view that the wavefunction is an abstraction, providing a mathematical reflection of certain aspects of reality, but not a one-to-one mapping.[241]

He further argues that the ontological cause is to be discovered at lower levels, below quantum mechanical dimensions, and that "quantum theory is inconsistent with the assumption of hidden causal variables" governing processes at the lowest, sub-quantum levels.[242] At the end of his Introduction to the 1993 book, *The Undivided Universe,* Bohm predicts the failure of the "current laws of physics" at the Planck length:

> The idea is that there will be a stochastic sub-quantum and sub-relativistic level in which the current laws of physics will fail. This will probably first be encountered near the Planck length of 10^{-33} cm. . . . This idea is connected with our ontological interpretation by means of a model of a particle as a sequence of incoming and outgoing waves, with successive waves very close to each other . . . . One of the main new ideas implied by this approach is that the geometry and the dynamics have to be in the same framework, i.e. that of the implicate order.[243]

This leads us to the observation that there exist three sets of framework-dependent mathematical models, applicable in three ranges: (1) Newtonian mechanics, operating at human dimensions, in the 10^{0} meter range, (2) quantum mechanics, operating at nuclear particle dimensions, in the 10^{-15} meter range, and (3) Bohm's "hidden variables," operating at the boundary of space–time, in the 10^{-33} cm range. Here Bohm reiterates an observation he had previously made regarding the important significance of the dimension 1.616 x 10^{-33} cm, the Planck length.[244] Over a decade earlier, in his 1980 publication, *Wholeness and the Implicate Order*, Bohm, in discussing the calculation of what is called "the zero-point energy" for point in space, provides another explanation for the significance of this 10^{-33} cm distance:

> We come to a certain length at which the measurement of space and time becomes totally indefinable. Beyond this, the whole notion of space and time as we know it would fade out, into something that is at present unspecifiable . . . . When this length is estimated it turns out to be about 10^{-33} cm. This is much shorter than anything thus far probed in physical experiments (which have gone down to about 10^{-17} cm or so). If one computes the amount of energy that would be in one cubic centimeter of space, with this shortest possible wavelength, it turns out to be very far beyond the total energy of all the matter in the known universe.[245]

It is here then, in the region of the Planck length at 10^{-33} cm, that Bohm predicts that there will be found *a boundary* separating an outer, *explicate order* from an inner, *implicate order*. There will also be seen gaps of, at minimum, 10^{-33} cm between each and every concentric shell that might be surrounding the central area, a spherical locus of Planck length diameter. This is in accord with the founding quantum theory of Max Planck (1858–1947), first published in 1900, in which he proposed that a photon could have only distinct, fixed energy states, separated by integers, which he termed "quanta."[246] Bohm uses such quantum theoretical considerations to point out a fallacy still held by many, the classical Cartesian assumption that space is continuous:

> What of the order between two points in space? The Cartesian order holds that space is continuous. Between any two points, no matter how close they lie, occur an infinite of other points. Between any two neighboring points in this infinity lies another infinity and so on. This notion of continuity is not compatible with the order of quantum theory . . . . Thus the physicist John Wheeler has suggested that, at very short distances, continuous space begins to break up into a foam-like structure. Thus the "order between" two points moves from the order of continuity to an order of a discontinuous foam.[247]

According to Bohm, at the ultimate bottom level of these subdivided infinities, between two points in space, will be found a cosmologically fixed boundary at the Planck length of 10^{-33} cm, and below this no further subdivision is meaningfully possible. It is here, at this smallest of possible spatial coordinates, that we encounter the origin of the dimension of space that finds its maximum in the diameter of the currently expanding universe (the bandwidth continually increases, at the speed of light).

Viewed another way, to use the depths of the ocean as an analog or metaphor, Bohm's sub-quantum mechanics can be viewed as being at the bottom of a dimensional ocean with a floor significantly below the depths of the region mapped and explored by the mathematics of quantum mechanics. The image in Fig. 9 plots three ranges of dimensional scale along a vertical axis.

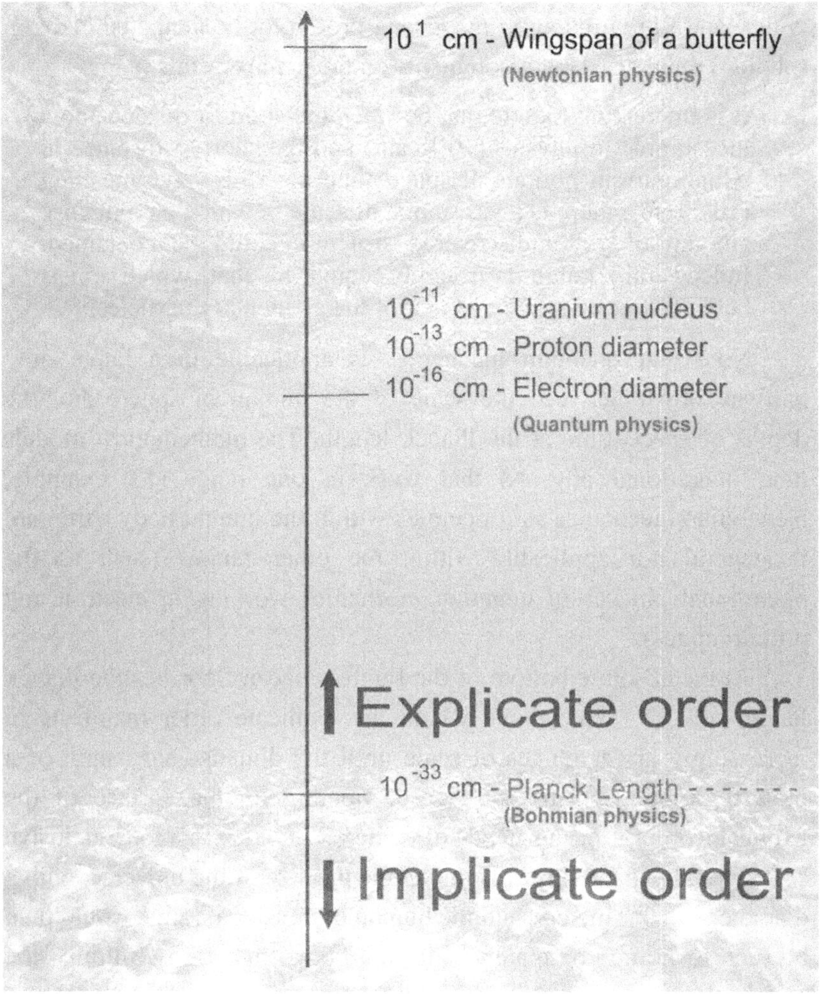

Figure 9. *Sub-quantum ocean floor metaphor.*[248]

On the same vertical axis can be seen dimensional ranges within which operate (1) classical Newtonian physics, (2) Quantum physics, and (3) Bohmian physics, each operating within distinct ranges, separated by over 10^{15} orders of magnitude. In a real sense, Bohm was plumbing this depth of dimensional physics, far below the level at which conventional quantum theory is applied, at depths of scale considerably below those of the dimensions of a Monarch butterfly wingspan, a uranium atom nucleus, a proton, or a calculated electron (photon) diameter. Bohm comments on the depths of this range:

> It is interesting to note that between the shortest distance now measurable in physics (10^{-16} cm) and the shortest distance in which current notions of space–time probably have meaning (10^{-33} cm), there is a vast range of scale in which an immense amount of yet undiscovered structure could be contained. Indeed this range is roughly equal to that which exists between our own size and that of the elementary particles.[249]

Note that quantum mechanics is applicable in a range only halfway down between the scale of the human biosphere and the depths of space–time at the Planck length. The mathematical models (and operational physics) that work in one range (for example, Newtonian mechanics as it operates within the human body parts) are in general not applicable within the other ranges (such as the operational physics of quantum mechanics working in electron and proton ranges).

At the absolute bottom of the figure's ocean floor lies the Planck length of 10^{-33} cm., above which the explicate order manifests in increasingly larger ranges of scale until the dimensional length of a butterfly (about 1 cm) is reached, floating on the surface of the explicate order's ocean depth of scale. Of course, a more inclusive scale would expand upwards to include the size of the universe, with a diameter of 10^{26} meters, and the human dimensional range would then be seen to lie approximately half way between these two limits, the size of the universe (10^{26} m) and the Planck length (10^{-35} m).[250] The complete scale spans the entire cosmos as depicted in Fig. 10. It should be noted that the Planck constant discussed in this book may be seen in two different forms depending upon whether meters or centimeters are the units of length under discussion, (i.e. The value 10^{-35} m is equivalent to the value of 10^{-33} cm., as centimeters are one-onehundredth of a meter)

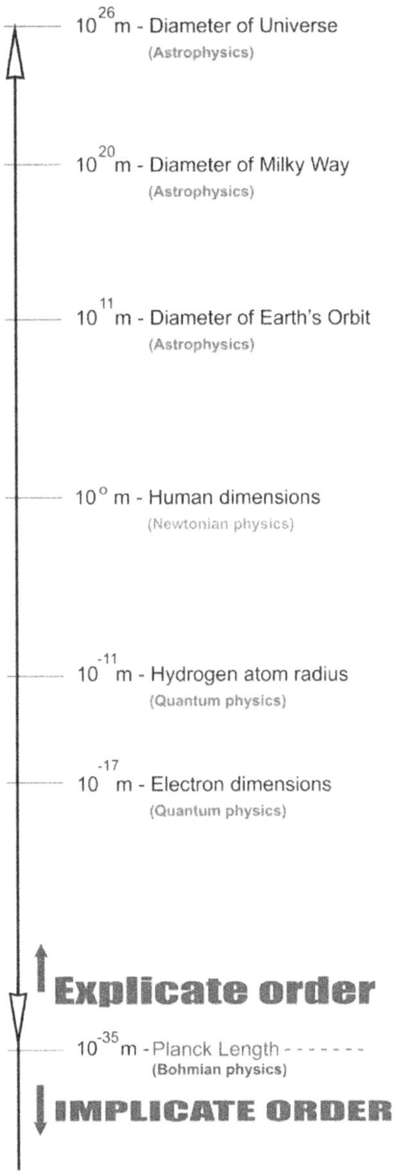

Figure 10. *Spectral domain and space–time domain.*

Another foundation upon which Bohm's model of the universe grew can be seen in his vision of an endless infinity of ordered fields, as he describes here:

> We may suppose that the universe, which includes the whole of existence, contains not only all the fields that are now known, but also an indefinitely large set of further fields that are unknown and indeed may never be known in their totality. Recalling that the essential qualities of fields exist only in their movement we propose to call this ground the *holomovement.* It follows that ultimately everything in the explicate order of common experience arises from the holomovement. Whatever persists with a constant form is sustained as the unfoldment of a recurrent and stable pattern, which is constantly being renewed by enfoldment, and constantly being dissolved by unfoldment. When the renewal ceases the form vanishes.[251]

Fields need not exist only within space–time. Einstein's theory of general relativity posits ten fields to describe the movement of gravity in four dimensions.[252] As early as 1919 a German mathematician, Theodor Kaluza, sent Einstein a paper proposing an extremely small fifth dimension in which additional frequency and phase fields could provide a solution to resolve Einstein's struggle to resolve the connection between gravity and electromagnetism.[253] In 1926 the Swedish physicist Oskar Klein "drawing on quantum mechanics, calculated the size of this compact dimension, arriving at a number that was tiny indeed—around 10^{-30} cm in circumference."[254] The theory became known as the Kaluza-Klein theory, offering the potential of extra fields in an additional dimension as a promising approach to resolving Einstein's search to connect gravity and electromagnetism. Einstein wrote to Kaluza, saying he liked the idea "enormously" and in fact continued working with the theory for the next 20 years; ultimately, however, the Kaluza-Klein theory was discarded because of its prediction of a particle so small that it could not be proven to exist using any foreseeable technology.[255]

But by the same token it has also been impossible to *disprove* a cosmological reality existing at such small dimensions, and it is here, at the limiting space–time scale of 10^{-33} cm, that Bohm hypothesizes an implicate order.[256] In Fig. 11, Pribram's diagrammatic concept of the Fourier transform is depicted with Bohm's implicate order to the left, in the spectral domain, and Bohm's explicate order to the right, in the space–time domain.

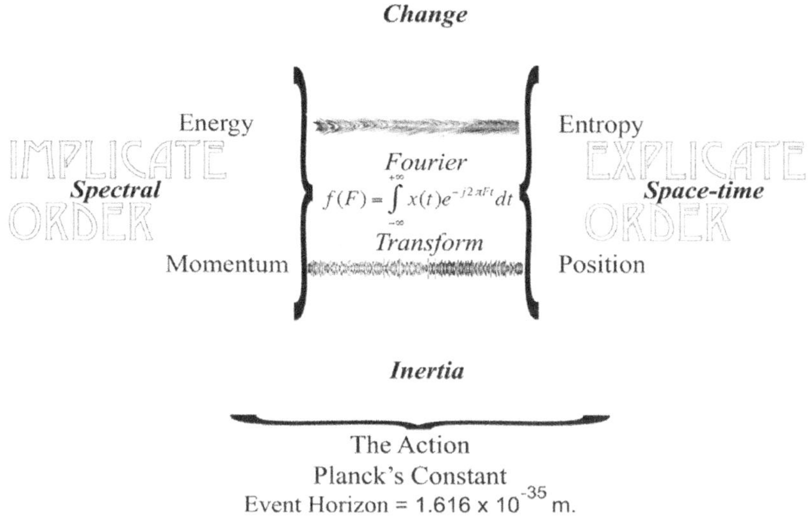

Figure 11. *The implicate/explicate order transform.*

In the diagram, the process linking the two orders is shown as the Fourier transform function, implementing a reciprocal flow of spectral information between the implicate order into the explicate order, the aforementioned two-way process, operating at an event horizon of the Planck limit of 10^{-33} cm, a process which Bohm terms the "holomovement," and which, he tells us, is grounded in the implicate order:

> The implicate order has its ground in the holomovement which is vast, rich, and in a state of unending flux of enfoldment and unfoldment, with laws most of which are only vaguely known, and which may even be ultimately unknowable in their totality . . . . Nevertheless, the overall law (holonomy) may be assumed to be such that in a certain sub-order, within the whole set of the implicate order, there is a totality of forms that have an approximate kind of recurrence, stability, and separability. Evidently these forms are capable of appearing as the relatively solid, tangible, and stable elements that make up our "manifest world."[257]

Holomovement and Holoflux

In 1978 Bohm discusses these concepts with remarkable clarity during a Harvard conference on "Science and Mysticism: Exploring the New Realities." After his presentation, which he called, "The Implicate Order: The Holodynamic Model of the Universe," Bohm was interviewed by Renée Weber for *ReVision* magazine, and here he expresses the difference between holography and the holomovement, and introduces the term "holoflux":

> **Weber**: Could we begin by clarifying the difference between the holomovement, the holograph and the implicate order?
>
> **Bohm:** Holomovement is a combination of a Greek and Latin word and a similar word would be holokinesis or, still better, *holoflux*, because "movement" implies motion from place to place, whereas flux does not. So the *holoflux* includes the ultimately flowing nature of what is, and of that which forms therein. The holograph, on the other hand, is merely a static recording of movement, like a photograph: an abstraction from the holomovement. We therefore cannot regard the holograph as anything very basic, since it is merely a way of displaying the holomovement, which latter is, however, the ground of everything, of all that is. The implicate order is the one in which the holomovement takes place, an order that both enfolds and unfolds. Things are unfolded in the implicate order, and that order cannot be entirely expressed in an explicate fashion. Therefore, in this approach, we are not able to go beyond the holomovement or the *holoflux* (the Greek word might be holorhesis, I suppose) although that does not imply that this is the end of the matter.[258]

In a theory congruent with Pribram's, Bohm hypothesizes that consciousness must operate dynamically in a holonomic fashion, driven by a quantum potential wave, and that the information guiding motion in space–time must originate from an *other order*, the implicate order, which lies outside of the commonly perceived explicate order of the space–time domain.[259] From early in his career, Bohm had tried to visualize the motion of quantum "particles," not moving in linear sequential tracks as viewed in a cloud chamber, but as oscillating in and out of space–time.[260]

Bohm, in a 1990 interview recorded for Dutch television, not only describes this movement or "holomovement" as superseding the so-called "particle" phenomena of quantum mechanics, but in doing so, he emphasizes the reality of the whole, the continuum of mind and matter, and discusses the general implications of this for perception:

> **Bohm**: Now, I saw that you could understand the quantum mechanics in terms of that process. That instead of saying an electron, for example, is a particle just moving along, you could think that there is a wave coming in, enfolding, or it's really unfolding at a point, it's enfolded in the whole universe. Then it folds back. Then another wave comes in, at a slightly different point. You get a series of points that are very close together so we imagine they're a particle, right? But, because of that wave nature, from which it comes, you understand, the wave/particle nature, that is, it's a wave.[261]

Support for this view is found in an interpretation of holodynamics put forth by a consciousness researcher at UC Irvine, Gordon Globus, who describes Bohm's holodynamic process as follows:

> There are two primary phases of this holodynamics according to David Bohm: *implication* and *explication* . . . . Implication is the "enfolding" of world to the whole and explication is the cotemporaneous "unfolding" of world from the whole. Each moment of unfolding/enfolding has a brief duration . . . . The world is explicate order continually unfolded from implicate order, rather than explicate order persisting.[262]

And what then might be the temporal rate of such an oscillation or "brief duration" of unfolding of the world in space–time? Here again it is the Planck constant that sets the limiting boundary condition. The Planck second, or "Planck time," calculated to be 5.44×10^{-44} seconds, is the smallest conceivable time interval consistent with quantum theory; below this interval, just as no further subdivision is meaningfully possible below the cosmologically fixed spatial boundary at the Planck length of 10^{-33} cm, time has no meaning. If indeed time is granulated at such a limit, then the cosmos may be created, destroyed, and recreated a staggering number of times per "human second." In the limiting case, space–time reality can be seen to emerge from an implicate order into an explicate order, from flux into space–time, at a recurring clock cycle every 5.44×10^{-44} seconds, only to collapse out space-time and back into the implicate order in alternate intervals between each of these sub-quantum clock cycles. Such a cyclic pattern of unfolding potential accords well with Bohm's description of the holomovement here, in a 1990 paper:

> All things found in the unfolded, explicate order emerge from the holomovement in which they are enfolded as *potentialities*, and ultimately they fall back into it. They endure only for some time, and while they last, their existence is sustained in a constant process of unfoldment and re-enfoldment, which gives rise to their relatively stable and independent forms in the explicate order.[263]

Meaning, Form, and Information

In developing an appreciation of the ontological and cosmological implications of quantum theory, Bohm placed great importance in maintaining a clear distinction regarding three terms: meaning, form, and information.

He describes the essential significance of form as being the conveyor and shipping container of embedded meaning:

> What is essential for a form to constitute information is that it shall have a meaning . . . . The form in the radio wave thus literally "informs" the energy in the received . . . this form is then *transformed* ("form carried across") into related forms of sound and light.[264]

How then is this "meaning" unpacked from its "shipping container?" How is information related to meaning? Bohm says that for information to become meaning requires, specifically, what he terms "active information."[265]

> The basic idea of active information is that a *form,* having very little energy, enters into and directs a much greater energy. This notion of an original energy form acting to "inform," or put form into, a much larger energy has significant applications in many areas beyond quantum theory.[266]

For Bohm, active information is the state in which a receiving energy is informed by the *form* of a transmitted energy; the receiving energy is "in-formed." Here he gives as an example a radio wave of electromagnetic energy being "in-formed"

> In the radio wave, the form is initially inactive, but as the form enters into the electrical energy of the receiver, we may say, that the information becomes active. In general, this information is only potentially active in the radio wave, but it becomes actually active only when and where there is a receiver which can respond to it with its "own energy."[267]

Bohm goes on to say that the operation of active information can be detected in nature outside of the human context, and can be seen, for example in the action of RNA as it decodes information locked in the form of DNA:

> it is assumed that the DNA molecule constitutes a code (i.e. a language), and that the RNA molecules "read" this code, and are thus in effect "informed" as to what kind of proteins they are to make.[268]

Finally, he identifies the *activity* of the energy in the receiving entity as the *meaning*, (i.e., the active information flow *is* the meaning):

> I would like to suggest then that the activity, virtual or actual, in the energy and in the soma *is* the meaning of the information, rather than to say that the information affects an entity called the mind, which in turn operates somehow on the matter of the body.[269]

What Bohm is implying here is that perhaps there is, in actuality, no "thing" that can be called mind, that there is no "entity called the mind," but that, rather, it is the *active information* itself, sourced in the implicate domain, that is in-forming the region of the explicit domain, directly. He concludes:

> The thinking process unfolds certain explicate thoughts, then folds back into the implicate, and then the next thought appears, the entire process forming a series of thoughts, which, seeming not too different, appear to be flowing in a continuous stream.[270]

Active Information

To better understand "active information," imagine that from each Planck length diameter "point" or "holosphere" in the universe (but outside of space–time) radiating waves of information-encoded energy emerge from the implicate domain at the center, radiating outward into space–time in expanding shells. Active information, like conceptual thought, does not need to be of great magnitude to be the effective cause of relatively large orders of magnitude of effect in the explicate domain. An example given by Bohm is the action of a tiny electrical signal that guides the flight of a large ship, airplane, or missile in flight. Here he provides a common example of how, even in the explicate world, a source of active information, comprising relatively minute quantities of energy, can effectively influence a receiver operating at considerably higher energy levels:

> As an example, let us take a radio wave, whose form carries information representing either sound or pictures. The radio wave itself has very little energy. The receiver, however, has a much greater energy (e.g. from the power source). The structure of the radio is such that the form carried by the radio wave is imposed on the much greater energy of the receiver.[271]

Here he claims that it is not the magnitude of energy that is important for active information to be effective, but *only the form itself*, guiding larger sources of energy in the explicate system. Bohm developed his mathematics of active information by deriving a potential function directly from the Schrödinger equation in quantum mechanics, and designating it as *Q*, the "quantum potential":[272]

> As time went on, Bohm emphasized the role played by the quantum potential. The quantum potential—whose influence depends on its overall form and not on its strength—is totally unlike anything previously postulated in physics.[273]

In order to prove that there is *real* causation acting in quantum mechanics, rather than accepting the commonly held theory that everything is all probability and stochastic process, Bohm proposed that a quantum potential (*Q*) also acts upon quantum particles such as electrons and protons and states that this *Q* function expresses the activity of a new kind of implicate order. This implicate order is immensely more subtle than that of the original field, as well as more inclusive, in the sense that not only is the actual activity of the whole field enfolded in it, but also all its potentialities, along with the principles determining which of these shall become actual.[274]

Bohm derives the quantum potential mathematically by taking derivatives of quantum mechanical Hamiltonian models of particle motion to give the following equation for the quantum potential.[275]

$$Q = \frac{-h^2 \nabla^2 |\psi|^2}{2m |\psi|^2}$$

Equation 1. *Equation of the Quantum Potential.*[276]

In Bohm's equation of the quantum potential (*Q*) of active information, the function *ψ* is the quantum field or "wavefunction" derived from Schrödinger's equation, *h* is Planck's constant, and *m* is the mass of the electron or "particle."[277] The symbol ∇ is called a "curl," and this symbol indicates an infinitesimal rotation of multidimensional coordinates (usually three or more) around a common origin, in order to obtain the differential motion of the wave in space–time. Bohm here points out what appears to be of special importance:

> what is mathematically significant in the above equation is that the wavefunction is found in both the numerator and the denominator . . . . The fact that ψ is contained both in the numerator and the denominator for Q means that Q is unchanged when ψ is multiplied by an arbitrary constant. In other words, the quantum potential Q is *independent of the strength, or intensity,* of the quantum field but *depends only on its form*. This is a particularly surprising result.[278]

Of great significance for Bohm is that fact that *Q*, the quantum potential, is seen to be independent of either strength or signal energy magnitude. This provides support for his insistence on causality, by implying that there is a formal cause, independent of magnitude, below the postulated randomness that is otherwise implied by the probabilistic calculations of quantum mechanics, and here he concludes:

> "In the causal interpretation, the electron moves under its own energy, but the information in the *form* of the quantum wave directs the energy of the electron."[279]

This can be visualized in the figure below, (Fig. 12, "Active Information and the Quantum Wave"), in which the function *Q*, sourced in the implicate order, molds or in-forms the quantum wave ψ in the explicate order. Here the metaphor of waves influencing the course of a ship is applicable.

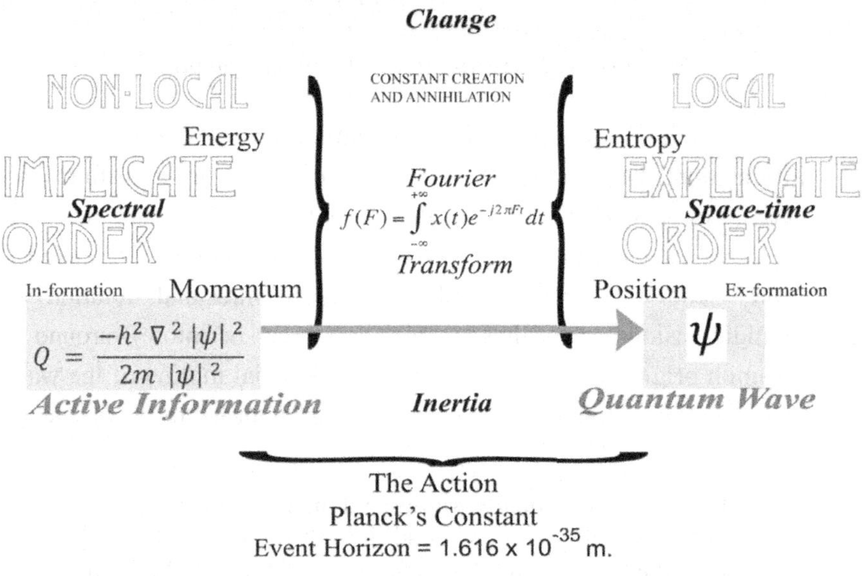

Figure 12. *Active Information and the Quantum Wave.*

In fact, a "pilot wave theory" similar to Bohm's active information model was proposed by de Broglie in 1927, though his idea was immediately criticized by the skeptical Wolfgang Pauli for being untestable.[280] In consequence, de Broglie abandoned his suggestion in favor of probability theory, but some years later Bohm brought up the idea again when he published, in 1952, "A Suggested

Interpretation of the Quantum Theory in Terms of 'Hidden Variables.'"[281]

In the figure, the Active Information function can be seen to be guiding the Quantum Wavefunction. Bohm here provides another example of *form* as active information in everyday human experience:

> It is clear, of course, that the notion of active information also applies directly to human experience. For example, when the form of a road sign is apprehended in the brain and nervous system, the form is immediately active as meaning.[282]

But not all information is active; Bohm refers to information which is not active as *virtual information*, "the activity of meaning may be only virtual, rather than actual. Virtual activity is more than a mere potentiality . . . it is a kind of suspended action."[283] Until it is made explicate as active information, this virtual information exists only in the implicate domain.

Indeed, Bohm's logic leads him to the conclusion that consciousness, at least of the nonactive kind, resides "within" the implicate order. This may be, of course, somewhat abstract, as the word "within" has no clear meaning for the spaceless, timeless implicate order. While Bohm speaks of the actual "substance" of consciousness as residing primarily within the implicate order, he sees the ranges of "human experience" of consciousness residing in reality as a whole, in a continuum of the implicate/explicate orders:

> In some sense, consciousness (which we take to include thought, feeling, desire, will, etc.) is to be comprehended in terms of the implicate order, along with reality as a whole . . . the actual "substance" of consciousness can be understood in terms of the notion that the implicate order is also its primary and immediate actuality.[284]

Bohm goes on to point out that some forms of information *from within* the implicate order are perceived, or can be perceived, by modes of consciousness in the explicate order: "Evidently this order is *active* in the sense that it continually flows into emotional, physical, and other responses." And here he provides as an example, music: "In listening to music, *one is therefore directly perceiving an implicate order.*"[285] However, it is not only in listening to music that consciousness resonates within the implicate order, but also in ordinary, everyday experience, as Bohm explains:

> A great deal of our difficulty comes from the fact that we accept the idea that not only matter, but all of our experience as well, is in the explicate order, and then suddenly we want to connect this up with consciousness, which is of a totally different order . . . we should consider the nature of ordinary, everyday experience. I say that it is totally misunderstood, that it is actually part of the implicate order . . . in that order everything contacts everything else and thus there is no intrinsic reason why the paranormal should be impossible.[286]

In a 1978 interview at the Krishnamurti center in Ojai, Bohm, who had been practicing meditation for over a decade, comments on distinctions between different modes of consciousness, and how it may be possible to experience the implicate order as a mode of consciousness which is timeless:

> **Question:** When mystics use the visualization of light they don't use it only as a metaphor, to them it seems to be a reality. Have they tapped into matter and energy at a level where time is absent?
>
> **Bohm:** It may well be. That's one way of looking at it. As I've suggested the mind has two-dimensional and three-dimensional modes of operation. It may be able to operate directly in the depths of the implicate order where this timeless state is the primary actuality. Then we could see the ordinary actuality as a secondary structure that emerges as an overtone on the primary structure . . . . The ordinary consciousness is one kind of music, and the other kind of consciousness is the other kind of music.[287]

The Sense of "I" and the Implicate Order

Bohm argues that none of this indicates that consciousness and the world of matter are absolutely separate, as assumed by Cartesian dualists. Consciousness and matter form a continuum, a whole, as he states here in the final words of his last book, *The Undivided Universe*, using the metaphor of a mirror:

> At no stage is there a break in this process . . . . There is no need, therefore, to regard the observer as basically separate from what he sees, nor to reduce him to an epiphenomenon of the objective process. Indeed, the notion of separateness is an abstraction and an approximation valid for only certain limited purposes. More broadly one could say that through the human being, *the universe is making a mirror to observe itself*. . . . All proposals are points of departure for exploration. Eventually some of them may be developed so far that we will take them to be actual working hypotheses.[288]

Here it may be useful to consider in greater detail the classical distinctions between two major metaphysical approaches to the universe: *dualism* and *monism.* Historically, most cultures have adopted one or the other of these two viewpoints, cherishing and defending one against the other as the very heart of their vision of reality. Bohm's approach to the cosmos as a "Whole" (e.g., in *Wholeness and the Implicate Order*), easily puts him on the side of monism, but Bohm's monism is a monism with an almost "trinitarian" flavor.

Monism traditionally takes one of two stances, "Physicalism," (MATTER > Mind), which argues the primary importance of matter in the scheme of creation. Physicalists see matter as primary, with mind as an emergent secondary principle, a mere epiphenomenon arising out of the many complex configurations of matter. Physicalism is the currently predominant paradigm of science in the early 21st century.

The second approach to monism is called "Idealism," (MIND > Matter), and here it is mind that is the primary underlying reality behind everything. This viewpoint was deeply explored by Immanuel Kant in the 18th century.[289] In idealism it is mind, the ideation process, that precedes matter, and perhaps projects matter, into the universe. To followers of transcendental idealism, it is the "idea" that has greater ontological reality than the thing being "ideated."

However there remains a third approach to monism, one in line with Bohm's sub-quantum physics and in accord with his concept of the implicate order. This third stance of monism is referred to as "Neutral Monism," (3rd SUBSTANCE > Matter & Mind), and this is

Bohm's understanding of the universe. In Neutral Monism, matter and mind are *both* taken to be derivative. The locus of matter and mind is to be found in space–time, while the more fundamental ontological reality is the "3rd substance," (lower right in Fig. 13), which is called by Bohm, the implicate order. Here, it is the implicate order that is primary, and matter and mind, projected resonantly into space-time, are derivative.

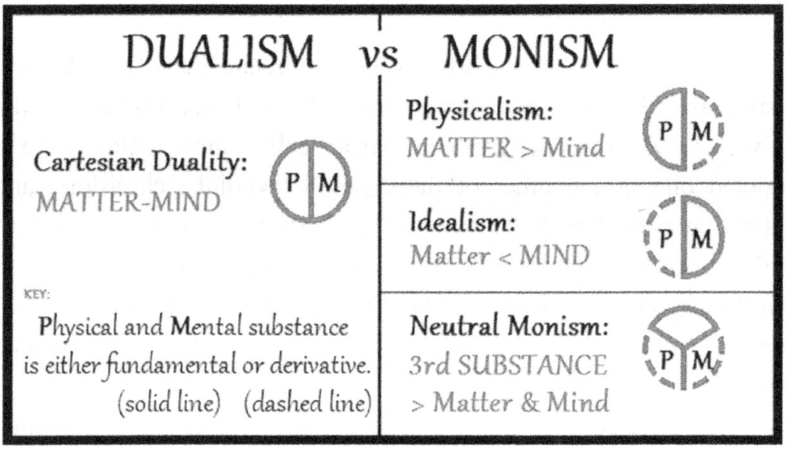

Figure 13. *Dualism vs. monism.*[290]

In an interview videotaped in Amsterdam in September of 1990, Bohm discusses the phenomenon of the sense of Self, that sense that individuals have of being "I," and in his interview he states clearly that this sense of self is a *product of thought*, and not the real situation. Rather, he points to the implicate order as being the *source* of a deeper reality, as being the true "I," the real "Me" which we all assume (and sense directly) to be so very unique:

> **Bohm**: Remember, thought is conditioned reflex at a very high subtle level. It just goes by itself, but it has in it the thought that thought is being produced by a center, which it calls "Me." And all the feelings, which should belong to that center, are thrown onto Consciousness as if from "Me," right?
>
> **Interviewer**: And our society reinforces that.

Bohm: Everybody says the same thing, like everybody sees the same rainbow, everybody sees the same "Self," whether it's there or not. In fact, it becomes all-important, the concept is regarded as central and supremely important. All sorts of powerful feelings will arise if it's questioned. You see, now, it all goes on by itself...and, you see... ...there's a deeper being, I say, which can...which is...that being which may be able to reveal itself anew...rather than being a fixed being...it's deep in the Implicate Order...in the Infinite.

Interviewer: So, thought creates "the Me," rather than "Me" creating the thought?

Bohm: Yes. If you don't see that happening, you will treat that as real and give it apparent reality. The whole being will take actions on that basis which seems to be coming from the center.[291]

Bohm's Quantum and Consciousness

Bohm's interest, like Pribram's, also lay in trying to understand and map consciousness and its role in cosmology, an area outside of the normal range of inquiry in both their respective professional careers. Bohm was forthright in his inclusion of consciousness as an important aspect of his quantum cosmology. According to Bohm there is a two-phase reciprocal movement or holomovement at the boundary of (or *between*) the implicate order and the explicate order,[292] and in a 1978 interview, he states unequivocally, "Let me propose that consciousness is basically in the implicate order."[293]

In the closing words of his final published work, Bohm reifies his proposal that consciousness and quantum mechanics are connected, that "they have the implicate order in common."[294]

> Several physicists have already suggested that quantum mechanics and consciousness are closely related and that the understanding of the quantum formalism requires that ultimately we bring in consciousness in some role or other . . . the intuition that consciousness and quantum theory are in some sense related seems to be a good one . . . . Our proposal in this regard is that the basic relationship of

quantum theory and consciousness is that they have the implicate order in common.[295]

Bohm's Waves About a Point

David Bohm saw a similar process underlying what proponents of quantum mechanics theorized to be simply "quantum jump" of energy. Here, in an interview with Bill Angelos in 1990, he describes the reality as waves coming in to a point and waves coming out of a point, for all points in the holoplenum:

Angelos: Could we… Again, the term "quantum" itself has its base in?

Bohm: Well, the "quantum"… That's one of the features of "quantum"… The quantum has its base in the fact that energy is transferred in discrete jumps or "quanta" rather than continuously…

Angelos: I see.

Bohm: Now, you can see that some sort of quantum appears here… the wave comes to a point, then there is a jump to another point. See one thing—according to quantum theory—was that the electron could go from one state to another without passing in between. Now you said: That is an utter mystery, right? But you see, if the wave comes in to this point… then it spreads out… it could come in to another point. So, therefore, it needn't look so mysterious.[296]

To more fully grasp the relationship of consciousness to the implicate order and the explicate order, two visual metaphors are now considered: the metaphor of the micro black hole, and the metaphor of the Yin/Yang symbol.

Metaphor of the Quantum Black Hole

It is appropriate here to use the image of a quantum black hole as a visual metaphor for realizing the topology of the implicate order.

> Micro black holes, which have also been called quantum mechanical black holes, quantum black holes, or mini black holes, are hypothetical tiny black holes . . . . It is possible that such quantum primordial black holes were created in the

high-density environment of the early Universe (or Big Bang).[297]

Visualizing the implicate order as a micro black hole, we can perhaps view its boundary as a spherical event-horizon shell, marking the limits of an "external," explicate order and the outer limits of an "internal," implicate order. Below the surface of such a sphere, of diameter equal to the Planck length, lies a transcendental implicate order, while outside the quantum black hole begins a universe of the explicate order, replete with time and space.

With this image in mind, it is appropriate here to review the topological understanding of a "point." The conventional definition of a point is a mathematical one, that it has no diameter at all, or that it is a sort of sphere with an infinitely small diameter. But a cosmological understanding of the point, considering that below the Planck length there is no space as we know it, the closest approximation to a point in space–time would seem to be a micro black hole of Planck length diameter.

Metaphor of the Yin/Yang Symbol

Bohm's cotemporaneous enfolding/unfolding process can be visually captured in the Yin/Yang symbol (Fig. 14), attributed historically to Chinese Taoist sages of the third century BCE.[298]

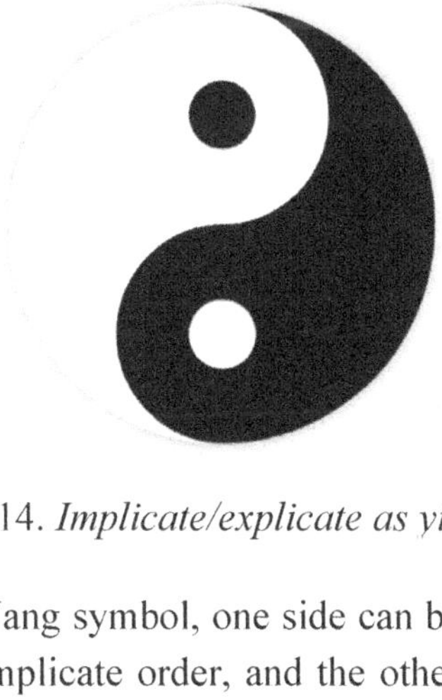

Figure 14. *Implicate/explicate as yin/yang.*

Using the Yin/Yang symbol, one side can be taken to represent a timeless, spaceless implicate order, and the other side can be seen as the expressed space–time explicate order. And yet each one is also at the heart of the other (designated by the dots in the two regions of the symbol).

It is of interest to note that the physicist Niels Bohr, famous for his visionary conception of electrons moving in fixed orbits around the nucleus of an atom, selected this Taoist symbol for his personal seal to adorn his "coat of arms" on the occasion of receiving the Royal Danish Award for Physics (Fig. 15).[299]

Figure 15. *Niels Bohr and his coat of arms.*[300]

That an ancient Taoist symbol might capture so well the essence of a cosmological process that theoretical quantum physicists have only arrived at in the twentieth century speaks well for the efficacy of human consciousness in previous ages and cultures for perceiving the mechanics of the cosmos via shamanic, contemplative experience.

Chapter 4: The Holoflux Theory

"Words strain,
Crack and sometimes break, under the burden,
Under the tension, slip, slide, perish,
Decay with imprecision, will not stay in place,
Will not stay still."[301]

–T. S. Eliot, "Burnt Norton," (1935)

These words from Eliot's *Four Quartets* seemed appropriate for introducing this chapter, in which an effort is made to use words to describe the holoflux theory, an abstract topology which integrate the theories of Bohm and Pribram. That words skillfully used can articulate even the most mathematical abstractions such as holograms and frequency domains can be seen in the following words from *The Holographic Universe*, in which Michael Talbot (1953-1992) succinctly express the implications of the Pribram-Bohm hypotheses:

> Considered together, Bohm and Pribram's theories provide a profound new way of looking at the world: Our brains mathematically construct objective reality by interpreting frequencies that are ultimately projections from another dimension, a deeper order of existence that is beyond both space and time: The brain is a hologram enfolded in a holographic universe.[302]

In this chapter, the integral ontological platform, initiated by Sri Aurobindo and Gebser, and carried forward through the work of Pribram and Bohm is extended through integration with properties of the electromagnetic field implicated by Pribram's approach to holonomic brain theory. The holoflux field theory of consciousness is more fully elaborated here as a process which can be approached visually as the nondual information-energy of awareness transforming in and out of two domains: an explicate order of space-time and an implicate, transcendent, nondual order that begins below the Planck length event horizon at the center, everywhere.

Figure 16. *David Bohm and Karl Pribram in Prague, June 1992.*[303]
Photo by Katherine Neville. Reprinted with permission.

That David Bohm and Karl Pribram influenced one another, and eventually became both professional and personal friends has been noted. In Fig. 16 can be seen the two enjoying themselves in Prague, with Bohm's wife, Saral, just four months before Bohm's untimely death in October 1992.

This theory is supported by a set of concepts from physics, electrical engineering, and mathematics. These concepts (bulleted below) are developed in this chapter through use of verbal description, analogy, and topological diagrams, with minimal recourse to equations or mathematical notation. In each case, focus is upon the ontological meaning of each concept:

- the Planck constant
- the holoplenum
- Wheeler's "Self-observing U"
- frequencies

- wavelengths
- Wien's law
- information
- superposition
- resonance
- dimensional analysis

The Planck Constant

Bohm makes frequent reference to a fundamental numerical value, the Planck length of 10^{-33} cm (or in terms of meters, 10^{-35}), a value widely used in quantum mechanical calculations, and derived from "the Planck constant" (denoted in calculations as $\hbar$). In speaking on the implicate order, Bohm states:

> It must be remembered that even this vast sea of cosmic energy takes into account only what happens on a scale larger than the critical length of 10^{-33} cm . . . . But this length is only a certain kind of limit on the applicability of ordinary notions of space and time. To suppose that there is nothing beyond this limit at all would indeed be quite arbitrary. Rather, it is very probable that beyond it lies a further domain.[304]

In 1900 the Berlin physicist Max Planck (Fig. 17), like Bohm decades later, found himself dissatisfied with what he saw as a problem with the tendency toward over-reliance on mathematical statistics; Planck himself was more interested in discerning the mechanism which would explain the underlying reality of the phenomenon of energy and thermodynamics.

Figure 17. *Portrait of Max Planck, 1878.*[305]

In his 1879 doctoral thesis, Planck discussed two major approaches to the second law of thermodynamics that were at the forefront of the physics of his day: one theory stressed the use of mathematically continuous equations to model unbroken flows of radiant field energy, while the other theory, the "atomist" theory, held that since matter is not continuous, it is preferable to apply statistics to calculate the mechanical motions of atoms and molecules in thermodynamics processes.[306] Planck felt that the atomist approach, relying as it did on statistical calculations, was masking the underlying realities. Over the next 20 years his research focused on the search for a nonstatistical mathematical model to reflect the ontological realities that linked radiation and the phenomenon of "heat."

On December 14, 1900, a date "widely acknowledged to be the date on which the quantum revolution began," Planck presented his derivation of a new radiation equation to a meeting of the German Physical Society, and in his presentation stated, "We therefore regard—and this is the most essential point of the entire calculation—energy to be composed of a very definite number of equal finite packages."[307] He went on to describe the constant in the equation which he had derived (later to be called the "Planck constant"), and stated that, "since it has the dimensions of a product of energy and time, I called it the elementary quantum of action." Planck postulated that electromagnetic energy both absorbed and emitted only in these elementary quanta of actions, or discrete packets of energy, which he also referred to as "energy elements." Planck's new theory, and his constant, published a month later in a 1901 issue of the journal *Annalen der Physik,* soon became a cornerstone of quantum mechanics.[308]

Earlier, in 1893, William Wien had discovered a mathematical relationship that allowed him to calculate the peak frequency of electromagnetic radiation for a given temperature, but his equations were only accurate for extremely high temperatures. Planck's paper solved the problem by assuming that any measurable electromagnetic fluctuation for a given temperature must consist of a unique spectrum of discrete frequencies, an integral set of separately calculable single frequency oscillators, each one making its own radiant contribution to the whole. Using his derived constant, Planck was able to calculate precisely the energy, $\boldsymbol{E}$, radiating in all directions from a theoretical "point" for a given macro "temperature."[309] The accepted value of this constant, as determined by Planck, is $\hbar = 6.62606957 \times 10^{-34}$ Joule-seconds.[310]

The Planck constant was derived specifically from other fundamentally agreed upon physical constants, including gravity and the speed of light. In presenting his constant, Planck expressed the importance of this constant in rather mystical terms, claiming that the discovery paved the way for, as he described:

> the possibility of establishing units of length, mass, time and temperature which are independent of specific bodies or materials and which necessarily maintain their meaning for all

time and for all civilizations, even those which are extraterrestrial and nonhuman, constants which therefore can be called "fundamental physical units of measurement."[311]

In his 1901 paper, Planck presented a simple equation to describe the energy output of a single "energy element" or "quantum." The equation is:

$$\varepsilon = \mathbf{h}\nu$$

In this equation the first term on the left, ε, the energy element, is equal to the product of Planck's constant times thefrequency, ν. The use of the constant **h** to unify, or normalize, the basic cosmological constants in physical calculations was immediately found to be so useful to simplify previously complex calculations, that the number was soon referred to as "Planck's constant," although in reality his discovery involved a set of constants, several of which are listed in Table 2.[312]

Table 2

Several Plank Constants

Constant	Value
Length	1.616199×10^{-35} meter
Time	5.39106×10^{-44} seconds
Area	2.6121×10^{-70} square meter
Power	3.62831×10^{52} watts

Note. Data adapted from Benenson et al. (2006), Handbook of Physics.

In the spring of 1905, a 26-year-old named Albert Einstein published four papers, soon to become famous in the world of physics. His first paper extended Planck's theory to include light, by asserting that Planck's radiation law only made sense if radiant energy of all wavelengths consists of discrete packets of energy, which Einstein called "light-quanta," or "photons."[313]

To obtain a better feel for the magnitude and use of these constants, let us calculate the velocity of one photon as it travels the distance of one Planck length in one Planck second. The equation for velocity is: velocity = length / time. If we substitute the Planck values for length and time (values found in Table 2), the velocity of a photon at the Planck level can be calculated as follows:

The equation:

Velocity = Planck Length / Planck Time

The values used:

Planck Length =	1.616199×10^{-35} meter
Planck Time =	5.391060×10^{-44} seconds

Substituting into the equation for Velocity:

Velocity = 1.616199×10^{-35} meter / 5.39106×10^{-44} s

or

Velocity = 2.99792×10^{8} meters / second

. . . which happens to be the measured constant speed of light in vacuum! If we try to move faster by decreasing the time it takes to move the same distance, the denominator in the velocity equation would have to be a smaller value, and a smaller value in the denominator would make the equation produce a higher velocity, violating the ceiling to the speed of light in a vacuum. Conversely, if we keep the same time interval and try to increase the distance travelled to something greater than the Planck length, we would again be increasing the velocity, exceeding the speed of light. Accordingly, there can exist no measurable (or conceivable) space smaller than that of the Planck length, and there can be no time interval smaller than that of the Planck second.

The Planck "second," then, at roughly 10^{-44} human "seconds," is the smallest possible time measurement, according to quantum mechanics, and is understood as a time period below which time has no meaning.[314] According to contemporary models of cosmological origin, this number also happens to be the time marked as the very beginning of our universe, at which time expansion into space began:

> The early inflationary universe is proposed to have had an accelerated expansion, the duration of which depends on the particular inflationary model. In any case, inflation is believed to have lasted an extremely short time: in the range of 10^{-33}–10^{-35} sec. In particular, in the chaotic inflationary model, inflation starts at the Planck time, $t_P = 10^{-44}$ sec.[315]

Similarly, the Planck length (at 10^{-35} meters, or its equivalent, 10^{-33} cm) is the smallest measurable length in physically defined space, below which space as we know it has no meaning. Below the Planck length one cannot say "smaller," because there is no spatial dimension definable below 10^{-35} meters. It is worth considering the contrast, between the Planck length of at 10^{-35} meters, and the diameter of a proton at 10^{-15} meters. The proton is enormous at an order of 20 magnitudes larger than a sphere of Planck length diameter.

It is also of interest to examine the last entry in the list of Planck constants shown in Table 2, the "Power" value, which indicates that a photon vibrating at the dimensional frequency corresponding to one Planck wavelength contains unimaginable power. The raw power flowing out from all of the stars in the known universe has been estimated to be 2 x 10^{49} watts.[316] The Planck Power, by contrast, has been calculated to be an incredible 10^{52} watts, three orders of magnitude *greater* than the luminosity of all of the stars in the visible universe combined.[317]

While quantum physics indicates that below the limiting Planck length there can be no space–time as we know it, Bohm predicts that it is precisely there, below the inner space–time boundary, that manifests a nonspatial, nontemporal domain which he identifies as "the implicate order."[318] The relationship of explicate order to implicate order should not necessarily be understood as a one-to-one relationship. On the contrary, as Bohm conjectures here, in a 1987 paper, there may in fact be *multiple explicate orders* (supporting the widely held "multiverse hypothesis"), suborders of a single, omni-connected implicate order:

> The principles of organization of such an implicate order can even define a unique explicate order, as a particular and distinguished sub-order, in which all the elements are relatively independent and externally related. To put it differently, the explicate order itself may be obtainable from the implicate order as a special and determinate sub-order that is contained within.[319]

The Sub-Quantum Holosphere

We have seen that Bohm proposes two orders, an explicate (space–time) order and what he terms "an implicate order beyond the critical limit of 10^{-33} cm," a sub-quantum (below space–time) order.[320] Having established the relative location of these two domains, it is useful to further identify the structural relationship of the implicate order to the explicate order by considering another image. The previous metaphor (Fig. 9) was presented in the image-metaphor of an "ocean depth," visualized in a vertical "up/down" geometry, having a "sub-quantum bottom" at the Planck depth of 10^{-33} cm. This ocean depth metaphor can be improved with a new metaphor based upon cosmology, that of a three-dimensional "outer/inner" shell of spherical geometry.

In the following diagram (Fig. 18, "A Sub-Quantum Holosphere"), the implicate order is seen as having its origin at the center of space, beginning at a spherical shell with a Planck length diameter of 10^{-33} cm, manifesting a spherical region that the physicist Paola Zizzi calls a "fuzzy quantum black hole."[321] According to Zizzi, "its event horizon is a fuzzy sphere."[322] Here, within the deepest "inside" possible in a space–time universe and located at the geometric center of every point in the space–time universe, lies the implicate order. In the figure, the sub-quantum holosphere or implicate order is depicted at the center of a spherical explicate region of the space–time universe, at the boundary separating time from the nontemporal. The holosphere is fixed and does not move, while around it move waves of space–time energy. The previous scale from Fig. 9 is shown here as a radius extending outward from the implicate order within the central sub-quantum holosphere.

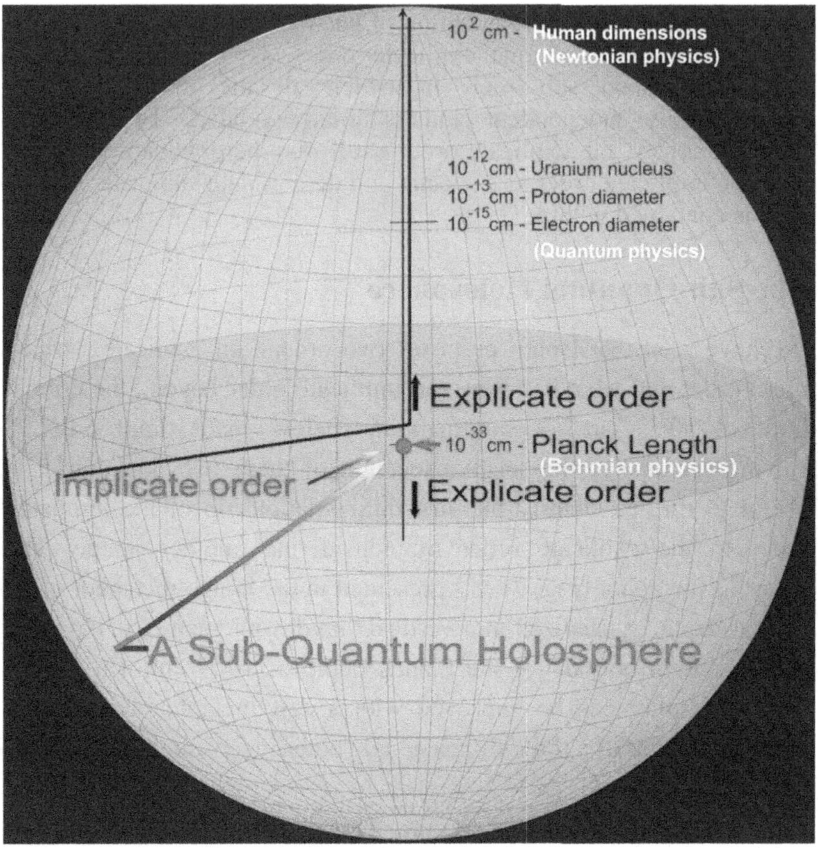

Figure 18. *A sub-quantum holosphere.*[323]

Again, the words of the poet T. S. Eliot are appropriate:

> And do not call it fixity,
> Where past and future are gathered.
> Neither movement from nor towards,
> Neither ascent nor decline.
> Except for the point, the still point,
> There would be no dance, and there is only the dance.[324]

The Sub-Quantum Holoplenum

This concept of a sub-quantum holosphere is mirrored both in Bohm's conceptualization of the implicate order as well as in contemporary theories of quantum computing, where quantum black holes are seen to carry and store information in the event horizon, described here by Paola Zizzi, a theoretical physicist at the University of Pavia, Italy:

> The event horizon of such a quantum black hole is the surface of a fuzzy sphere . . . . Fuzzy quantum black holes and quantum human minds can be viewed as subroutines of the whole quantum-computing universe . . . . This cosmic network exploits quantum communication complexity, and, thanks to entanglement, allows its parts to accomplish a distributed task without the need of any kind of communication. This is possible because all its parts share a common quantum space–time background, which is nonlocal.[325]

The quantum physicist John Archibald Wheeler (1911–2008) has hypothesized that a considerable amount of information is storable on the surface boundary of a spherical black hole. His theory is discussed in detail elsewhere in this chapter, but here it is worth mentioning that his calculations indicate that the surface boundary of a one-centimeter diameter spherical black hole is capable of storing 10^{64} bits of information.[326] Compared even to the total number of stars in the universe, which has been estimated to be 10^{24}, this is a great deal of potentially storable information.[327]

Assuming the existence of a sub-quantum holosphere, the next question to ask would be how many subquantum holospheres might there be, underlying our universe? Perhaps to coincide with Big Bang theory there was only one, out of which the universe of space–time emerged. An alternate approach would be to imagine the universe as consisting of an infinity of quantum black holes, each one having a diameter of the Planck length. In such a model, the implicate order would be distributed evenly, to be found at the center, everywhere, throughout the space–time cosmos. Bohm describes such a model, arguing here for a cosmological bedrock of actuality, which he calls "the plenum."[328]

> It is being suggested here, then, that what we perceive through the senses as empty space is actually the plenum, which is the ground for the existence of everything, including ourselves. The things that appear to our senses are derivative forms and their true meaning can be seen only when we consider the

> plenum, in which they are generated and sustained, and into which they must ultimately vanish.[329]

Bohm points out that in early Greece, the philosophers Parmenides and Zeno also held that "space is a plenum."[330] Here, located in a communion of Planck holosphere, we find a plausible explanation for the peculiar observed phenomenon that the universe, from every point, appears to be expanding outwardly, directed radially *away from each point*; thus, in the space–time universe, the center is everywhere:

> Relativity assumes that the universe will appear much the same no matter where the observer happens to be situated. The nebulae appear to be rushing away from our particular position, and they will appear to be rushing away from any other position in which an observer is located.[331]

Thus we hypothesize this Bohmian holospheric plenum, like some sea-floor bed of existence, to be discoverable at the center of every conceivable coordinate in space–time; the concept can be seen resonating vibrantly in the title of Jean Gebser's magnum opus, *The Ever-Present Origin*.[332] We can find the same image resonating in T. S. Eliot's "still point of the turning world."[333]

> O hidden under the dove's wing,
> hidden in the turtle's breast,
> Under the palm tree at noon,
> under the running water
> At the still point of the turning world.
> O hidden.[334]
>
> –T. S. Eliot, "Coriolan," (1931)

We also hypothesize that this center of the cosmos, the origin, located everywhere throughout the cosmos, is the same origin that Teilhard defines as "point Omega."[335] Having established the ontological basis of a plenum, Bohm goes further to envision what he terms "the holomovement," the dynamics of energy processing within this plenum, an "immense sea of energy," flowing within and between the non-time-space domains of the implicate and the explicate orders:

> This sea is to be understood in terms of a multidimensional implicate order, while the universe of matter as we generally observe it is to be treated as a comparatively small pattern of excitation. This excitation pattern is relatively autonomous and gives rise to approximately recurrent, stable and separable projections into a three-dimensional explicate order of manifestation, which is more or less equivalent to that of space as we commonly experience it.[336]

This region of implicate order, the holoplenum (Fig. 19), is here visualized as consisting of close-packed holospheres, each a sphere bounded by the Planck limit of 10^{-33} cm.

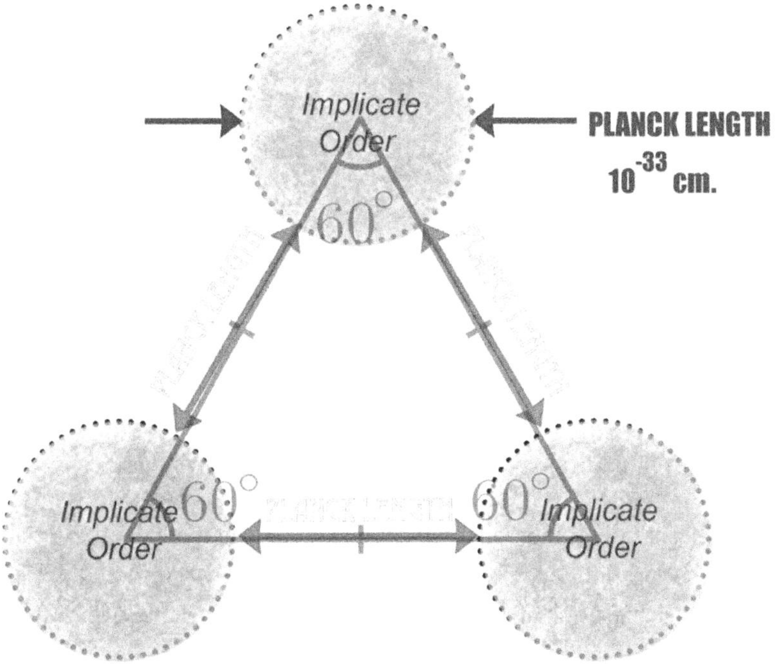

Figure 19. *Holoplenum formed by implicate order.*

Because each holosphere can come no closer than one Planck length to neighboring holospheres, simple geometry insures that in a single plane these holospheres form equilateral triangles, and in three dimensions the holospheres of the holoplenum would thus form regular tetrahedrons in close-packed configuration (Fig. 20). This geometry aligns with the prediction of the architect and systems theorist Buckminster Fuller, who wrote that the natural analytic geometry of the universe is based on arrays of tetrahedra; Fuller supported his conjecture with the fact that the strongest possible homogeneous truss is cyclically tetrahedral.[337]

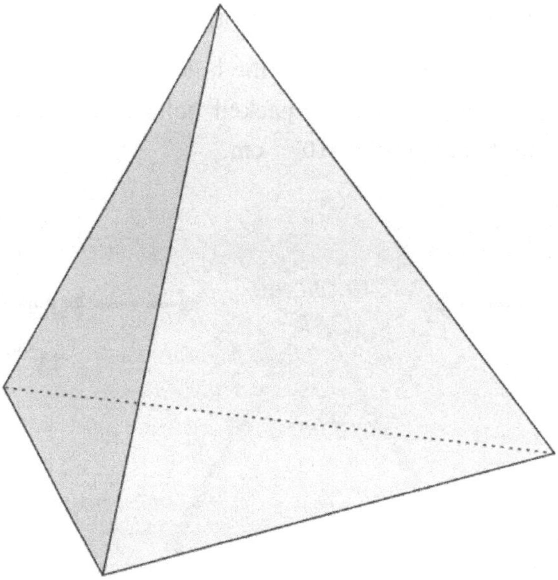

Figure 20. *A regular three-dimensional tetrahedron.*

Fuller's idea, that a three-dimensional tetrahedron may be a natural underlying geometry of nature, has been reinforced by the discovery, in 2013, that DNA strands at the nanostructure level form stable three-dimensional tetrahedrons and therefore are suitably scalable for nanotechnology engineering:

> The physicochemical stability of DNA makes it a desirable candidate for the design of nanostructure materials as DNA is inherently nanoscalable . . . . The DNA nanostructure consists of six single-stranded oligonucleotides that hybridize to form a *three-dimensional tetrahedron*.[338]

In addition to operating within a geometrically maximally stable plenum, any phenomena at the Planck dimension would also operate at the highest frequency rate possible in space–time, simply because at this smallest dimension would also be found the smallest wavelength, both being equal to one Planck length. According to Wien's law (the higher the frequency, the higher the energy), at the smallest wavelength possible, at the Planck length, would also be found the highest energy.[339]

Wheeler's Self-Observing U

> "Truth is less than truth until it is made known."[340]
>
> –John Archibald Wheeler

In the context of consciousness, another visual metaphor is revealing in its congruency with Bohm's cosmology. The metaphor can be seen in a relational diagram drawn by Wheeler, who had worked closely with Niels Bohr in the late 1930s. Wheeler later became a Princeton colleague of Bohm's in the late 1940s, and was the thesis advisor to the young Richard Feynman, who himself later received the Nobel Prize for Physics in 1965.[341]

Wheeler used a visual metaphor (Fig. 21), which he termed the "Self-Observing U," to help understand the quantum theoretical observation and creation of an actual entity, within an explicate (space–time) universe. As Wheeler explains here in his 1998 memoir, *Geons, Black Holes, and Quantum Foam,* he views the universe as a self-excited circuit:

> Not all potentiality is converted to actuality in any finite time . . . . My diagram of a big U (for universe) attempts to illustrate this idea. The upper left end of the U represents the Big Bang, when it all started. Moving down along the thin left leg and up along the thick right leg of the U symbolically traces the evolution of the universe, from small to large—time enough for life and mind to develop. At the upper right of the U sits, finally, the eye of the observer. By looking back, by observing what happened in the earliest days of the universe, we give reality to those days. The eye could as well be a piece of mica. It need not be part of an intelligent being. The point

> is that the universe is a grand synthesis, putting itself together all the time as a whole.[342]

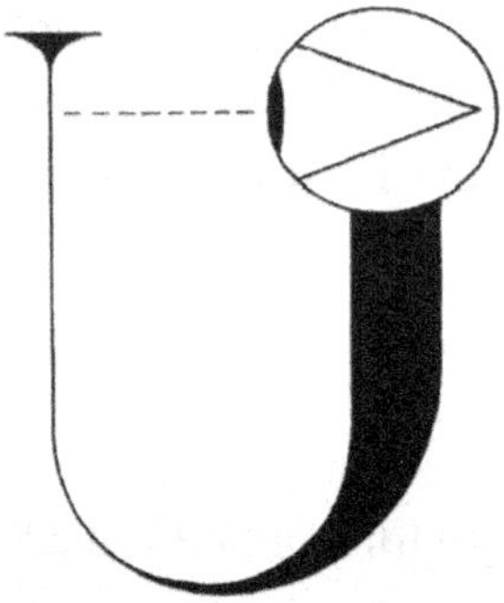

Figure 21. *Wheeler's self-observing U.*[343]

The figure also supports Bohm's conjecture that there is no separation between the observing instrument and what is observed, as he elaborates here:

> For example, you are looking at this table; the form of this table has been built up by your experience which you are projecting into the table. Is the table you, or is it something separate from you? You appreciate it as a table with a certain form and a certain subsistence, but that form and that subsistence are as much you as the table. If you probed it with a very high energy machine, say with neutrinos, they would go right through, and the table would be a vaporous nebula. So the form of the table as a solid substance, or subsistence, comes from the human brain with its own particular mode of interacting. In a sense, the observer is the observed.[344]

That this paragraph on perception sounds a great deal like passages from Karl Pribram's 1972 *Languages of the Brain* is likely no coincidence, as Bohm had previously met Pribram at a Krishnamurti conference in Ojai, and the two had become personal acquaintances and shared many ideas over the preceding five years.[345]

Creation and Dissolution

In the same sense that a digital computer system has a clock-cycle to which all events are synchronized, what might be the underlying clock-cycle of the universe? According to quantum theory, the clock-cycle can be no less than the Planck second, for, as previously discussed, any time unit smaller than Planck time has no meaning. As Bohm's colleague, the physicist Hiley states:

> The explicate "moments" that unfold have a certain temporal thickness, a certain duration. Moments are temporal segmentations of unfoldment . . . . This is "a moment where *what has been* is separated from *what is yet to come*." Within this brief duration time becomes nonlocal.[346]

Thus there is separation, and a cycling between local and nonlocal time, and this can be no less than the Planck length. Figure 22 illustrates such a hypothetical clock-cycle with a Planck-limited value of 5.39106 x 10^{-44} seconds as the clock cycles the cosmos on and off, in and out, forming the digital holoflux.

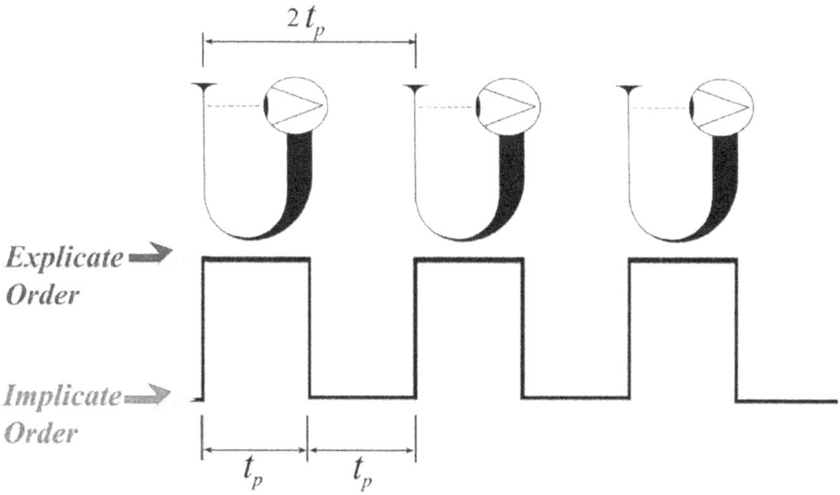

t_p = Planck time (10^{-44} seconds)

Figure 22. *Clock-cycle: digital holoflux.*

This theoretical explicate/implicate digital clock cycle is supported in quantum theory by the concept of "quantum fluctuations," described here by Wheeler:

> Space—pure, empty, energy-free space—all the time and everywhere experiences so-called quantum fluctuations at a fantastically small scale of time, of the order of 10^{-44} second and less. During these quantum fluctuations, pairs of particles appear for an instant from the emptiness of space—perhaps an

> electron and an antielectron pair or a proton and an antiproton pair. Particle–antiparticle pairs are in effect all the time and everywhere being created and destroyed. Their destruction is so rapid that the particles never come into evidence at any everyday scale of observation.[347]

This accords well with Bohm's ontological model, in that each quantum tick reveals a new explicate order:

> As with consciousness, each moment has a certain explicate order, and in addition it enfolds all the others, though in its own way. So the relationship of each moment in the whole to all the others is implied by its total content: the way in which it "holds" all the others enfolded within it.[348]

Here we are reminded of the power of the words of the poet T.S. Eliot, where this holoflux clock-cycle is intuited in the opening stanza of one of his greatest poems:

> Time present and time past
> Are both perhaps present in time future
> And time future contained in time past.
> If all time is eternally present
> All time is unredeemable.
> What might have been is an abstraction
> Remaining a perpetual possibility
> Only in a world of speculation.
> What might have been and what has been
> Point to one end, which is always present.[349]
>
> –T. S. Eliot, "The Four Quartets," (1943)

The holoflux clock-cycle process or oscillation between an implicate order and an explicate order is also remarkably congruent with Dewey's previously-mentioned theory of laminated space–time.[350] Going further, in considering that, within an explicate space–time order, any frequencies in time imply the existence of corresponding frequencies in space, Bohm's comments:

> These ideas are connected with our ontological interpretation by means of a model of a particle as a sequence of incoming

> and outgoing waves, with *successive waves very close to each other.*[351]

These "very close" waves find their successive limit, as we have discussed, at the bounding Planck length of 10^{-33} cm, and the bounding Planck time value of 10^{-44} sec time can also be viewed as a total of 10^{44} successive alterations ("incoming and outgoing waves") per (human) second.

Temperature of a Quantum Black Hole

Looking at a calculation of the theoretical temperature of an "object" of the size of a holosphere which is assumed, for purposes of calculation, to be a black hole, we can use an equation derived by Stephen Hawking to calculate the temperature at the surface of such a quantum black hole (Equation 2, "Hawking's temperature of a black hole"):

> Notice that Hawking's formula, the mass of the black hole is in the denominator. That means that the larger the mass, the colder the black hole, and conversely, the smaller the mass, the warmer the black hole.[352]

$$T = \frac{1}{16\pi^2} \frac{c^3 h}{GMk}$$

Equation 2. *Hawking's temperature of a black hole.*[353]

According to the Hawking equation and Wien's law, we see that the highest temperature, the greatest energy and highest frequency of electromagnetic energy possible, all issue from the very heart of each and every "point" in space–time, from within the theoretical locus of a holosphere of diameter 10^{-35} m, the Planck length. Thus the quantum wave, or Planck wave of quantum number one, enters the space–time continuum at maximum temperature, from every holospheric point in the holoplenum, at a cyclic frequency of 10^{44} Hz, a frequency limit set by the Planck time of 5.39106×10^{-44} seconds.

Wheeler's "It from Bit"

Even beyond timing considerations, Wheeler also developed a theory of information storage, in which he finds at the upper limit the Planck length, deep within the physical world:

> "It from bit" symbolizes the idea that every item of the physical world has at bottom, at a very deep bottom, in most instances an immaterial source and explanation.[354]

Wheeler's explanation was to digitize the cosmos by assigning one "bit" of information to the smallest physical area possible in space–time: an area equal to that of a square having each side the Planck's limiting length of 10^{-35} meters.[355]

In 1970 one of Wheeler's graduate students, Jacob Bekenstein, proposed a novel idea, that there is a maximum amount of information that can be stored in a finite region of space, and that quantum theory points to that area:[356]

> One unit of entropy (information), one unit of randomness, one unit of disorder, Bekenstein explained to me, must be associated with a bit of area of this order of magnitude . . . . Thus one unit of entropy is associated with each 1.04×10^{-69} square meters of the horizon of a black hole.[357]

This proposed upper limit to the information that can be contained upon the surface of a specific, finite volume of space has come to be known as the "Bekenstein bound."[358] Figure 23 is Wheeler's depiction of information bits stored on the surface of a sphere.

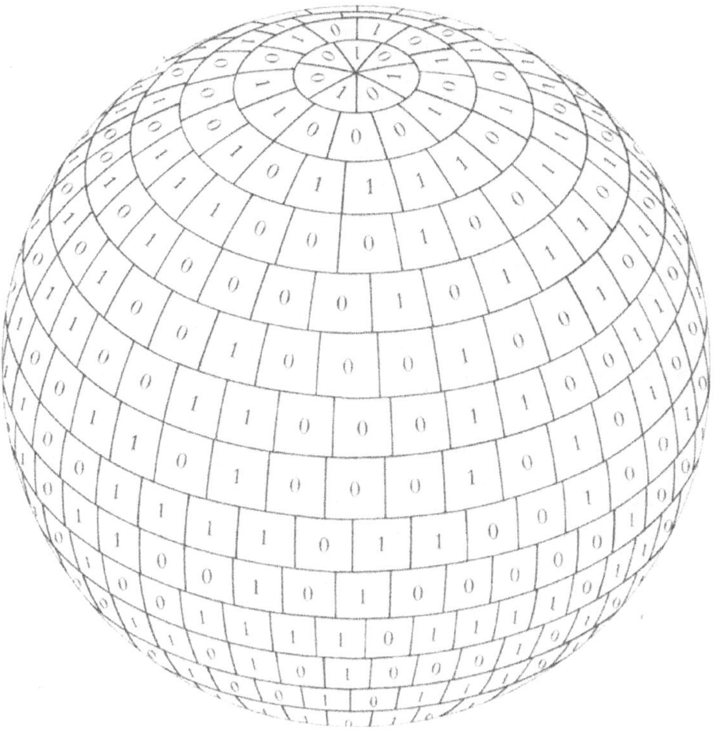

Figure 23. *Bekenstein information bits.*[359]

Using this concept, we can calculate the maximum amount of information that might be stored in a one-centimeter radius sphere located in the human brain. To obtain this limiting number of bits, we simply calculate the surface area on a sphere of one-centimeter radius, and then divide the result by 1.04 x 10^{-69} square meters (a square of Planck length).

Using simple geometry, we use the equation for the surface area of a sphere $4\pi r^2$. We can then approximate the area to be:

$$4\pi r^2 = 4 \text{ x } 3.14 \text{ x } (10^{-2})^2 \text{ or } 1.256 \text{ x } 10^{-5} \text{ square meters.}$$

If we divide this value by 1.04 x 10^{-69} square meters (the square of the Planck length) we obtain the following answer for *the number of bits of information* stored in a one-centimeter radius sphere:

$$1.256 \times 10^{-5} / 1.04 \times 10^{-69} = 1.207 \times 10^{64} \text{ bits of information.}$$

By contrast, if we perform the same calculation using the surface area of a holosphere (quantum black hole), the surprising answer (perhaps not so surprising) is that there are π qubits (quantum "bits") per holospherical surface, which implies that the universe, in terms of information computation, is not based on two-state binary computing, but that it operates on π-based computational algorithms at the base level of the holoplenum.

The enormous amount of information that might be physically stored in the holoplenum accords well with the holographic principle described by Paola Zizzi, the proponent of a quantum gravity theory in which space is considered to be granular. According to Zizzi, the universe should be regarded as a kind of super computer:

> The holographic principle states that all the information enclosed in a region of space with volume *V*, is encoded on the surface ***S*****,** bounding *V*. More precisely, each pixel of area (a pixel here is one unit of the Planck area) of ***S*** encodes one bit of information. In a more sophisticated (quantum) version of the holographic principle the encoded information is quantum, and each pixel encodes one quantum bit (or "qubit") of information . . . . At this point every quantized horizon can be viewed as a quantum register (a quantum register is the memory of a quantum computer, built with qubits).[360]

Problem remain however. Just *how* might this information be stored, and more importantly, in order to address Chalmers' "hard problem," what might be the mechanism for comprehension of this information? How is such information experienced? To answer "how" for any of these problems first requires a clear understanding of how information is encoded, transmitted, and received via twenty-first-century human technology. We begin with the concepts of frequency and the frequency domain.

Fields and the Frequency Domain

An understanding of fields and frequency, their measurement and dynamics, are of central importance to an understanding of the holoflux field theory of consciousness. It was Michael Faraday (1791–1867) who first conceived of the reality of an invisible electromagnetic "field" during his observation of magnetic lines of force around a conductor carrying a direct current.[361] But while Faraday had the ability to conceptualize a certain ontological reality, he did not have sufficient mathematical training to formalize the dynamics of his theorized field. Thus it was James Clerk Maxwell (1831–1879) who in 1861, through a series of interlocking equations, subsequently known as the "Maxwell equations," was able to model the dynamics of electric and magnetic energy, in waves of various frequencies, flowing through space–time in an ocean of vibrating flux.[362]

Figure 24. *James Clerk Maxwell at Cambridge.*[363]

His discovery was a mathematical one, based upon his development of a set of four interlocking partial differential equations with which he was able to predict the dynamics of electric charges and how they are generated by oscillating magnetic fields. Ironically, though Maxwell's equations predicted electromagnetic waves, their existence was not accepted as real until nearly 25 years later, when Heinrich Hertz (1857–1894) experimentally detected electromagnetic waves.[364] Nevertheless, Hertz was forever in awe of the equations that led him to his discovery, and here he speaks, with almost mystical reverence, of the "wisdom" that he has found in Maxwell's equations:

> One cannot escape the feeling that these mathematical formulae have an independent existence and an intelligence of

their own, that they are wiser than we are, wiser even than their discoverers, that we get more out of them than was originally put into them.[365]

In the 1800s, frequencies were measured in units of "Cycles Per Second," a convention which was adopted for many decades into the twentieth century. With the development of quantum mechanics, it was realized that units of frequency often needed to be dimensionless for certain calculations to make sense. In 1930, a dimensionless frequency unit named, appropriately, "Hertz," was adopted by the International Electrotechnical Commission, officially replacing the misleading concept implied in units of "cycles per second."[366] Karl Pribram, commenting on the confusion, here suggests using the term "spectral density" in lieu of the term "frequencies":

> When we deal with a spectrum of frequencies per se, we often substitute the term "spectral density" to avoid a common confusion that Gabor pointed out in his paper: non mathematicians, even physicists, often think of frequencies as occurring only in time, but when mathematicians deal with vibrations as in the Fourier transform, frequencies are described as composing spectra and devoid of space and time.[367]

Perhaps Pribram is reacting here to the confusion in terminology that has muddled some critics of Pribram's holographic paradigm, leading them to believe that frequency by definition (as units of cycles per second) can only exist in space–time. Here for example, Ken Wilber exhibits this fallacy, apparently misled by the term "cycles/second," in his dismissal of Pribram's holographic paradigm:

> The transform of "things" into "frequencies" is not a transform of space/time objects into space/time frequencies. Frequency does not mean "no space, no time"; it means cycles/second or space per time. To read the mathematics otherwise is more than a quantum leap; it is a leap of faith.[368]

Holoflux in The Holoplenum

Holoflux, as defined in this thesis, is a transformational process of modulated energy flowing into, out from, and within a holoplenum of holospheres in frequencies as high as 10^{44} Hz. This holoflux manifests a flowing field of consciousness, the modulation of which is information embedded in the waves and frequencies of space–time energy. Within each holosphere, below the Planck length of 10^{-35} meters—though the words "within" and "below" are clearly inappropriate for a region that is nonlocal—lies what Bohm calls the *implicate order*.

Holoflux can be regarded as a type of plasma energy that manifests consciousness in two different modes within two different domains: (1) In the space–time domain, where it manifests in electromagnetic waves, the holoflux process consists of reflective, time-sequenced snapshots of awareness, like glass beads on a thread, and (2) In the frequency domain of the nonlocal, the implicate order, holoflux manifests as immediate awareness (a timeless-sense, the "consciousness-without-an-object" reportedly experienced by the Harvard mathematician turned philosopher Franklin Merrell-Wolf).[369]

Holoflux, as a total manifestation, is a dual process of reflective consciousness and transcendent awareness. Communication, and consciousness flow between and within the two domains: the implicate and the explicate. Another way of regarding holoflux is to see it as a two-way moving flux: (1) It is an explicate energy (E) moving *inward* into the implicate order at the holospheric event horizon of a quantum black hole, and (2) Moving in the opposite direction, it is an implicate energy (Q) moving *outward* as quantum potential from the event horizon, as holoflux energy transforms from out of the implicate order into the explicate order. My thesis rests upon the assumptions that Bohm's implicate order is real, that the frequency domain is real, and that his active information wavefunction is real, affecting our world of "particles" as it springs from the virtually infinite nonlocal information holoflux within the implicate order.

Like Bohm, electrical engineers separate the world into two domains, which they term the *complex-frequency domain* and the *time domain*. It is our normally perceived space–time that electrical engineers identify as the *time domain*, but it is in the *frequency domain* that they see the operation of communication signals.[370]

Franklin F. Kuo, an American computer scientist and communications engineer at Bell Laboratories, wrote an electrical engineering textbook, developing the mathematics of communication in *Network Analysis and Synthesis*. In the first chapter of his classic textbook, Kuo describes the two domains, and the mathematical advantage of creating and receiving communication signals within the *frequency* domain as opposed to the *space–time* domain:

> In the *complex-frequency domain*, the voltage–current relationships for the elements are expressed in *algebraic* equations. Algebraic equations are, in most cases, more easily solved than differential equations. Herein lies the raison d'être for describing signals and networks in the *frequency domain* as well as in the *time domain*.[371]

These two orders of electrical engineering theory can be seen to reflect Bohm's explicate and implicate orders. Bohm's explicate order is associated with the engineering *time* dimension through which signals are transmitted and received sequentially in time, while Bohm's implicate order can be identified with the electrical engineering *frequency* dimension as a timeless, spaceless domain of pure frequency-phase information.

It has been discovered that these two domains are strongly linked, mathematically, with the operation of the Fourier transform. As Kuo says, "between time and frequency, the translation is effected by the Fourier series."[372] As we have seen in Chapter 4, the Fourier series has also become the bedrock of quantum theoretical calculations, as expressed by David Bohm in the opening of his textbook, *Quantum Theory*:

> "It seems impossible to develop quantum concepts extensively without Fourier analysis."[373]

Our thesis, then, is that Kuo's frequency domain, bridged to time by Fourier's mathematics in our twenty-first-century information technology, has an ontological reality, and that there does exist an actual frequency domain, which we also identify with Bohm's hypothesized implicate order.

The Dynamics of Holoflux

Having established the basic architecture of the holoflux theory, it is possible to go further by imagining the cyclic nature of holoflux information as it manifests in and between the two domains, from a Bohmian perspective.

During each inward (implicate) interval (of 10^{-44} seconds), Bohm's quantum potential undergoes reconfiguration calculations within the nonlocal frequency domain. These internal calculations use information that has been feeding inwardly from the explicate (space–time) order during previous "cycles." The sub-quantum calculations within the implicate yield solutions involving frequency superposition and fractal pattern formation, likely far beyond our current mathematical capabilities, within the frequency domain. The results of this holoflux superposition-calculation are then posted into space–time, and incorporated into the outflowing modulation of holoflux as the quantum potential *Q*, of "active information." These manifest in waveforms entering space–time as fields of modulated electromagnetic energy.

Conversely, information flowing within space–time, not only travels outward, but moves (in-folds and enfolds) towards the center, back into the implicate order, to be captured by the quantum black hole of the Planck holosphere at its event horizon. Such enfolding information is immediately available within the implicate order/frequency domain for superposition-calculation of the next out-going quantum wave into space–time. The resulting outgoing wave of quantum potential (*Q*) of "active information" is encoded with information to nudge every photon in the universe in space–time.

Holoplenum Plasma versus Flat-Screen

It is possible to comprehend the emerging projection of space–time reality from the holoplenum by expanding upon the metaphor of a flat-panel plasma display. The human visual threshold for detecting separate images lies somewhere between 10 to 12 separate images per second, and accordingly the industry standard in the motion picture industry is 24 frames per second, a standard which ensures that the presentation of a sequence of completely static images appear to the human observer as smooth and continuous motion.[374] By contrast, if the entire universe flashes in and out of existence at the limiting Planck time of 5.3×10^{-44} seconds, equivalent to a "frame" rate of almost 10^{44} "frames per second," the cosmos would obviously *appear* to be smooth and continuous in all respects, certainly to any human observer of the cosmos, even at quantum dimensions.[375]

The approximate image resolution of a "holoplenum display" obtained by dividing one inch by the Planck length, yields a maximum resolution of 1.584×10^{34} holopixels per inch. At such hyper-fine resolution, even a proton in the 10^{-15} meter dimensional range would appear to be moving smoothly through space.

Quantum Biology and the Holoflux Field

Quantum mechanics and the concepts of the holoplenum and the holoflux field provide tools for understanding the biological basis for human modes of consciousness.

The anesthesiologist and consciousness researcher Stuart Hameroff has, along with the British physicist Sir Roger Penrose, proposed a quantum theory of human consciousness based upon quantum field activity within microtubule structures within the human body.[376] It is notable that in a recent paper, Hameroff argues for an approach to consciousness studies in terms of quantum mechanics, and refers to his own contribution in terms of an emerging "quantum biology:"

> At the turn of the 20th century, physics was perceived by its practitioners to be as solid as rock, only to be shaken to the core by the discovery of quantum mechanics. Today, perhaps the most dynamically expanding branch of science is molecular and cellular biology, which is amassing impressive reams of data well ahead of its proper integration and deep analysis. Quantum biology is emerging gradually as a response to the challenge of explaining such important, yet poorly understood phenomena as photosynthesis, bioenergetics, vision, olfaction, bird navigation, etc. Yet, the grandest challenge of all is to explain how our brains work and, in particular, how does conscious behavior emerge from

the structure and function of the human brain and its cellular and sub-cellular components.[377]

Quantum Brain Dynamics

Quantum brain mechanics, and in particular, Quantum Brain Dynamics, a field pioneered by the physicists Hiroomi Umezawa and Herbert Fröhlich in the late 1960s, is one such contribution to Hameroff's proposed quantum biology.[378] As early as 1978, Stuart, Takahashi, and Umezawa proposed a mechanism of human memory and consciousness, consistent with quantum field theory, which they called "quantum brain dynamics (QBD)."[379]

Based on Umezawa's original research, Mari Jibu and Kunio Yasue at Seishin University in Japan put forward a detailed model of consciousness based upon the quantum dynamics of the electromagnetic field, which they describe as follows:

> The long-standing difficulty of understanding the origin and mechanism of unity of consciousness (as a unified self) has been called the "binding problem" in neurophysiology. The binding problem may be solved . . . by investigating the usually neglected quantum electromagnetic phenomena taking place in the dynamically ordered regions (i.e., perimembranous regions) of intracellular and extracellular water. There, each brain cell is enfolded within a common field of macroscopic condensation of evanescent photons and all the physico-chemical processes taking part in the stratified society of brain cells are subject to the control and unification by quantum electrodynamics. Unity of consciousness thus arises from the existence of the global field of condensed evanescent photons overlapping the whole brain tissue in the cranium.[380]

In extending Umezawa's theory of QBD, Jibu and Yasue describe how they have discovered that, through dynamic resonance (spin), polarized dipole water molecules throughout the blood stream produce the activity, in effect, of a single resonant water dipole, a "macromolecule" in the bodies of all living, water based creatures. The macromolecular-dipole, offering an exceptional information highway for electromagnetic fields on the macro level, manifests as a plasma. It should be noted that it was the physics of plasma that captivated David Bohm during his thesis research under Oppenheimer, and it is useful here to clarify the definition of plasma:

> A state of matter in which a significant portion of the particles are ionized. Plasmas are by far the most common phase of matter in the universe, both by mass and by volume. All stars are made of plasma, and interstellar space is filled with plasma. Common forms of plasma include lightning, St. Elmo's fire, the polar aurorae, the solar wind, neon signs, and plasma displays in modern home television.[381]

Within this polarized, nonlocal, H_2O quantum field is embedded an electromagnetic field described here by Jibu and Yasue:

> There are two basic fields in QBD: the water rotational field and the electromagnetic field. The two must be described simultaneously by quantum field theory because they interact with each other . . . . The water rotational field extending over the whole assembly of brain cells manifests an overall domain structure in which the whole region occupied by water is divided into many domains . . . . In other words, they are bound to each other quite strongly by a quantum theoretical nonlocal coherence.[382]

In 1995, building upon their earlier paradigm, Jibu and Yasue published the mathematics to support their theory that the creation and annihilation dynamics of "corticons" points to an "energy quanta of the water rotational field extending to the whole assembly of brain cells, and photons, that is, energy quanta of the electromagnetic field."[383] Their theory focuses upon the nonlocal quantum field that is generated within the human circulatory system, embedded in an electromagnetic field manifesting throughout the contiguous ionized , blood plasma system.

That our blood might act as an electromagnetic plasma should not be surprising— the first line in a textbook on plasma physics reads, "It has often been said that 99% of the matter in the universe is in the plasma state."[384] Jibu and Yasue go further by explaining how QBD explain how life itself can be seen to be equivalent to "the unity of water" in the human body, or as "a single molecule" in the human blood system:

> All H_2O molecules bound together in a QBD vacuum domain form a single, extensive molecule of water in a macroscopic domain. That is, water throughout the entire region of the cerebral cortex is thought to be composed of many macroscopic water molecules whose sizes are all comparable to the coherence length of the QBD vacuum, that is, about 50 microns. This remarkable feature of water in living matter might provide us with quantum field theoretical support for the idea that life is nothing but the unity of water as a single molecule in living matter.[385]

Quantum Brain Dynamics supports the model of a locus of consciousness beyond that of the nervous system and brain. The supporting pattern can be found in the QBD manifestation of a "single extensive molecule of water," dipolar in geometry like an antenna, generating a complex magnetic field through the circulation of hydrogen ions in blood plasma.

In such a model, the blood system of the human body can be seen as an extensively polarized "super cell" of nonlocal electromagnetic plasma energy, which can then be differentiated from the neuronal brain body of consciousness, itself generated by sequential electrical patterns flowing as pulses in the nervous system pathways. Moving charges generate magnetic fields, and ionized human blood flow is no exception: flowing blood plasma results in creation of a magnetic field, and this is in accord the conjecture of QBD.[386] The circulatory system can be seen as a magnetic plasma composed primarily of ionized red blood cells (referred to as erythrocytes) and water molecules, flowing together in complex vortices of blood plasma around every cell and through every capillary of the body.[387] Each erythrocyte is a flexible, annular, bio-concave disk shaped like a doughnut (in geometry, a torus), having a thin webbed center where the hole in a doughnut would be located (Fig. 25).

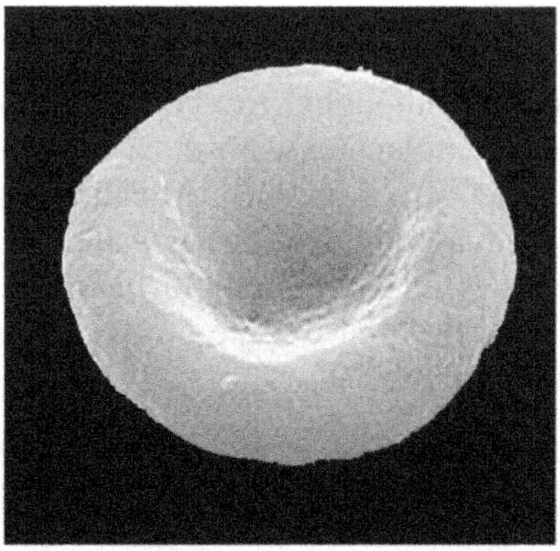

Figure 25. *Erythrocyte (Red Blood Cell).*[388]

Mammalian erythrocytes are unique among vertebrates for having no nucleus at all; each blood cell sheds its nucleus and internal organelles as it reaches maturity. The typical outside diameter of a red blood cell is approximately 9 microns, close to the infrared wavelength of 9.6 microns generated by the human body.[389]

The adult human body contains approximately 6 grams of iron, of which 60% is stored throughout the 10^{12} erythrocytes, each of which contains approximately 270 million atoms of ionic iron embedded in transparent hemoglobin.[390] Thus each erythrocyte, replete with iron ions embedded in hemoglobin, creates in effect, an ionized iron toroid.[391] One of the best-known applications of the toroid shape is called the "Tokamak device," which uses a magnetic field to confine plasma in the shape of a torus in experiments to produce controlled thermonuclear fusion power. Andrei Sakharov, who eventually became known internationally as a Soviet dissident, first designed the Tokamak enclosure device in the Soviet Union in the 1950s.[392]

Holoflux Analysis of a Red Blood Cell

Let us step through a holoflux analysis of the possible functions of a red blood cell as a locus of consciousness and the possible use of the erythrocyte as a locus of memory storage at the human biological scale. If, as previously conjectured, the red blood cell has an ideal diameter to resonate electromagnetic radiation in its ferrite-embedded ring at the human infrared wavelength of 10 microns, then we can ask if this configuration could accommodate a unique isospheric frequency (wavelength) for each of the currently 7 billion living humans on the planet. Figure 26 outlines the feasibility of this approach.

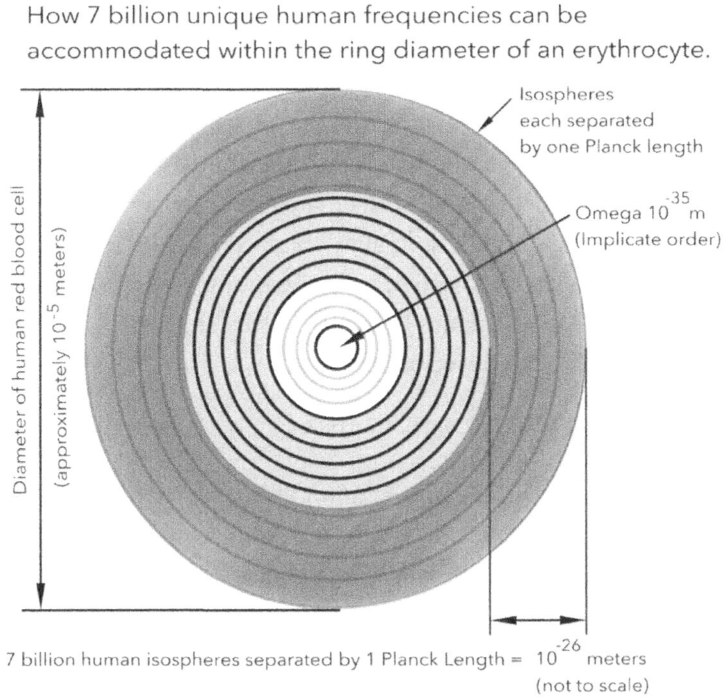

Figure 26. *Isospheric capacity of red blood cell.*

Assuming each unique frequency would match its radially unique isosphere, separated by only one Planck length, Figure 26 suggests how seven billion unique isospheres, each of quantum discrete frequency, might be nested within the geometry of a typical human red blood cell (in the image we have multiplied 7 billion times the Planck length of approximately 10^{-35} m to obtain an estimated shell thickness of 10^{-26} meters). This implies that each living human being might have a unique holospheric frequency, detectable by other human blood cells via the implicate order about which each is centered, and thus provides a possible mechanism for communication, via the mechanisms of resonance and superposition in the frequency domain of the implicate order.

A calculated estimate of the memory storage capacity of a single red blood cell can now be made, using Wheeler's approach to determine the number of Bekenstein equivalent bits on the surface of an isosphere the diameter of a typical red blood cell.[393] Here we take 8.1 microns, (or 8.1×10^{-5} meters), as the average diameter of a typical human red blood cell, and proceed to calculate the maximum possible storage capacity on the surface of a hypothetical holosphere associated with a single red blood.[394] To obtain the limiting number of bits, we simply calculate the surface area on a sphere of 8.1 microns diameter, and then divide the value obtained by 1.04×10^{-69} square meters (which is the Bekenstein unit of entropy, or approximately the square of the Planck length, 1.616199×10^{-35} meter).[395] The surface area of the erythrocyte-bisected sphere is:

$$4\pi r^2 \text{or } 4\pi(8.1 \times 10^{-5})^2 = 4\pi(6.561 \times 10^{-9}) = 8.24 \times 10^{-9} \text{ sq.meter.}$$

Dividing this surface area of an erythrocyte by the qubit area of 1.04×10^{-69} square meters yields the estimated maximum storage capacity for a typical human erythrocyte of *8×10^{60} bits of information.* This is an extremely large information storage capacity. By contrast the National Security Agency's Utah Data Center (officially known as the Intelligence Community Comprehensive National Cybersecurity Initiative Data Center) is estimated to have a maximum capacity, when completed, of a mere *1×10^{18} bits of information.*

Not to be neglected as part of any electromagnetic holoflux process associated with the cardiovascular system is the likely role of the heart:

> The heart generates the largest electromagnetic field in the body. The electrical field as measured in an electrocardiogram (ECG) is about 60 times greater in amplitude than the brain waves recorded in an electroencephalogram (EEG). The

magnetic component of the heart's field, which is around 5000 times stronger than that produced by the brain, is not impeded by tissues and can be measured several feet away from the body with Superconducting Quantum Interference Device (SQUID) based magnetometers.[396]

Dimensional Analysis

As previously discussed, a rule of thumb in applying the phenomenon of resonance in any situation indicates that the optimal physical dimensions of the resonant receiving system must approach a half-wavelength of the dimension of the transmitted energy wave. Using *dimensional analysis*, this fact can be used in the search for possible loci of consciousness in the electromagnetic spectrum. In this section we examine calculated correlations between dimensions of structural bio-components in the human body and corresponding wavelengths in the electromagnetic spectrum.

If we examine human biophysical components and calculate their peak resonant electromagnetic energy frequencies using Wien's Law, and then identify the frequency bandwidth within which we find these resonances for the biological components being examined, two possible systems emerge: the near infrared and the near ultraviolet (the use of the word "near" implies frequencies adjacent to the visible spectrum).

The following dimensional analysis correlations support the possibility of a conscious electromagnetic–holoflux information system active within a 10 micron (μm) wavelength band throughout the human bio-system.

- The human body at 98.6 F emits wavelengths which peak at an average value of 9.6 μm.

- Earth has an atmospheric pass-band at wavelengths of 9–10 μm.

- Inner diameters of blood capillaries are 10 μm.

- Outside diameter of a typical blood cells is 9 μm.

- CO_2 laser outputs can be tuned to emit between wavelengths of 9.4 –10.6 μm.[397]

Data is presented in Table 3, ranking physical components by increasing size, along with their calculated Wien blackbody peak wavelengths and corresponding theoretical waveguide resonant frequency ranges.

Table 3

Physical Correlations with Electromagnetic Frequency Bands

Physical component	Range (physical dimension)	Calculated blackbody wavelength peak	Calculated electromagnetic resonant waveguide frequency range
Passband of Earth's atmosphere	1 μm to 11 μm	3 to 50 μm = "Mid-Infrared Spectrum"	*10^{12}–10^{13} Hz*
Blood capillary system	10 μm	3 to 50 μm = "Mid-Infrared Spectrum"	*10^{12}–10^{13} Hz*
Red blood cell	6 μm to 8 μm	3 to 50 μm = "Mid-Infrared Spectrum"	*10^{12}–10^{13} Hz*
Cerebral neuron nucleus	3 μm to 18 μm	3 to 50 μm = "Mid-Infrared Spectrum"	*10^{12}–10^{13} Hz*
Unmyelinated cortical axons in the brain	80 nm to 400 nm	10 to 400 nm = "Ultraviolet Spectrum"	*10^{16}–10^{17} Hz*

Microtubule in neuron and in cilia	15 nm to 25 nm	10 nm to 400 nm is the "Ultraviolet Spectrum"	*10^{16}–10^{17} Hz*

Note: nm = nanometer (10^{-9} meter); Hz = Hertz.
Data adapted from *Romanes, Textbook of Anatomy.*

As indicated in Table 3, the average diameter of human blood capillaries of 10 μm coincides with a major "passband" of the planetary atmosphere itself. As previously discussed, a simple Wien's law calculation show that the normal human body radiates most strongly in the near infrared, measured at a macro level as "heat" at a temperature of 91.6 F, which again, using Wien's Law, exhibits a peak spectral energy response at wavelengths of approximately 10 μm. It is accordingly to this wavelength that the maximum sensitivity of military night-vision goggles is tuned.[398]

A physiological waveguide candidate in another frequency range can be found in the inner diameter of the microtubule system, located throughout neurons and also within cilia.[399] At 15 nm in dimension, microtubules would be especially sensitive to an electromagnetic wavelength in the ultraviolet bandwidth, just above visible light frequencies in the electromagnetic spectrum.

Biosphere, Magnetosphere, Chromosphere

In searching for energies of consciousness, it is important to explore the possibility of consciousness as having a component (if not comprising) the Earth's magnetic field, also known as the geomagnetic field, the magnetic field of energy flux extending from the Earth's inner core out to its interaction with that stream of high energy particles emanating from the Sun, the solar wind. This enormous magnetic field in which we find ourselves is generated by a magnetic dipole, powerfully driven by a spiraling ocean of liquid metal in the outer core, spinning around the enormous copper-magnesium crystal of the Earth's inner core. A supercomputer at the Los Alamos National Laboratory has created a model of flow patterns of geomagnetic energy within and around the Earth, illustrated below in the supercomputer generated model of Fig. 27.

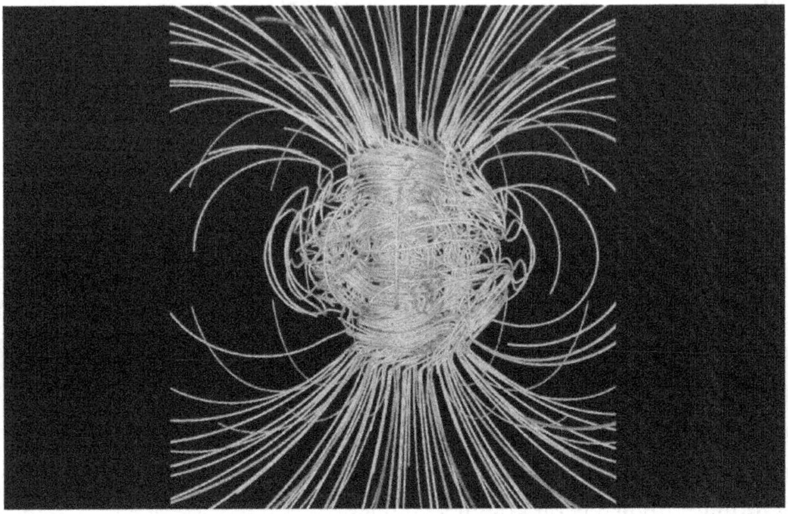

Figure 27. *Supercomputer model of magnetosphere.*[400]

Each line in the supercomputer model represents the locus of an intensely compressed magnetic field line. The greatest density of electromagnetic plasma flux is found in the clusters of complex magnetic lines below the crust extending throughout the Earth's inner core. The dipole axis of the Earth is tilted at an approximate angle of 11 degrees with respect to the rotational axis, as if there were a bar magnet placed at that angle along the central axis of the Earth.

However, unlike the static field of a bar magnet, Earth's geomagnetic field quivers and resonates over time in synch with the motion of molten iron alloys in the Earth's outer core; this phenomenon is referred to as the *geodynamo*, a phenomenon that is used to explain the presence of anomalously long-lived magnetic fields in astrophysical bodies.

If the universe is alive and sentient, then these vibrations could be replete with frequency modes of consciousness, communication signals yet to be deciphered by human technology, filling bandwidths that have perhaps only thus far been explored by mystics, shamans, and cybernauts.[401]

Associated with the magnetosphere are two fundamentally different categories of energy generated in the earth's thin crustal layer: *geothermal* and *geoelectric* energy. While we are all aware of geothermal energy spectacularly displayed in the glowing rivers of lava emerging from volcanic activity, what is not so widely known is that volcanic eruptions also generate enormous electrical discharges in what is termed *volcanic lightening* (Figure 28).

Figure 28. *Volcanic lightening discharges.*

Geoelectromagnetic volcanic lightening discharges are readily apparent at night in the rushing clouds of ash and gas rising high into the surrounding atmosphere. While the amount of *geothermal* energy in the world is estimated to be 2×10^7 Megawatt-hours (MWh), it is dwarfed by the amount of *geoelectric* energy, estimated at 1.14×10^8 MWh.[402]

Besides volcanic lightening, other major sources of geoelectric energy in the atmosphere include plasma generated by the globally continuous discharges of lightning in the upper atmosphere, as well as the shimmering curtains of ionized plasma in the colorful aurora borealis. All of these sources of planetary electromagnetic energy contribute, through superposition, to the composite geoelectric activity of the planet, and can be observed in the quivering of a single compass needle. Each human body is continually immersed within these fields.

Vast currents of electromagnetic energy flow within the salt water depths of the ocean as well as within conductive channels in the Earth's crust, forming enormous rivers of electricity.[403] These geoelectric flows are called *telluric currents* (from Latin *tellūs*, "earth"), and they occur both in the Earth's crust as well as within the much deeper mantle. These geomagnetically induced currents are generated by the continual interaction of the sun's magnetic field and the spinning Earth's geomagnetic field at the geopause, the area where the sun and the earth meet in ever-changing resonant magnetic vibrations of enormous power that induce the aurora borealis.[404]

Nikola Tesla (1856-1943), who invented the alternating current electrical grid system, was thought to have fabricated a system to tap Telluric currents for power. His plan was to use the currents themselves to transmit electrical power in wireless electrical energy grids across the planet, hoping to establish a system of free energy distribution over the face of the Earth, assisted by the planet's omnipresent geomagnetic field.[405] Mysteriously, his pilot experimental station which would have tapped the Telluric currents was burned to the ground prior to being brought on-line. As there was a bitter rivalry between Tesla and Thomas Edison, who attempted to make his direct current distribution system the standard, some say this may have been an act perpetrated by Edison's supporters or Edison himself.[406]

This enormous geoelectric energy, spread all over the earth's crust, is most concentrated in wet rocks located within zones known as earthquake belts.[407] If human consciousness does have an energy component in the electromagnetic spectrum, then it must be affected by, and interact to some degree with, this geoelectric energy environment of the earth. It has been discovered that animals possess a "magnetic sense" and appear to be able to track the magnetic field of the earth, and the corresponding flow of ions in the atmosphere.[408] Magnetoreception is an accepted phenomenon among a wide range of animals: birds, fruit flies, honeybees, turtles, lobsters, sharks, stingrays, whales and even bacteria.[409] Even animals not normally known for their migration habits have been discovered to possess such a sense:

> Recent publications from a German research group (Begall et al., 2008), made the discovery that cattle (and other herd animals, such as red and roe deer) tend to situate themselves on a magnetic North–South axis, as if involuntarily directed by the earth's magnetic field. These surprising results were discovered when satellite images provided by Google Earth were used to analyze herding patterns and behavior. However, the built-in magnetic compass gets out of alignment the closer the cattle get to high voltage power lines, and the cattle then align with the power lines instead.[410]

Contemporary research in electrophysiology indicates that human bodies may also be more able to sense or detect electromagnetic fields than has previously been acknowledged. Research at Cal Tech has discovered traces of magnetite in the human brain and heart, in about the same density as that found in the aforementioned migrating animals, and has proven that onset of rapid eye movement in sleeping humans is shortened in the E–W directional orientation of sleepers as compared to the N–S position.[411]

On the average there are from about one hundred lightning discharges per second around the globe.[412] It is known that one of the physical characteristics of the earth is that an effective waveguide cavity exists between the ionosphere and the surface of the earth, and this cavity resonates at 7.8 to 8.0 Hz, in what is called the Schumann resonance frequency peak.[413] It is, remarkably, the same frequency peak found in high end Theta waves and low end Alpha waves as measured on the human scalp.[414] In studying this Schumann cavity resonance phenomenon, the neurophysiologist Michael Persinger concludes:

> Because of the very small attenuation rates within the cavity, electromagnetic patterns are generated for megameters without appreciable loss of amplitude. In general, the expansion and return of the "ringing" around the earth's circumference can occur between 4 and 5 times. The latency for the return of the circular wave is about 125 milliseconds.[415]

These lightning discharges, within this geophysical cavity, emit energy waves that travel around the globe and return in 125 ms cycles. In one second, each wave circles the planet eight times; this, in effect results in an ongoing lightning driven pulse rate equal to the cranial wave frequency of 8 Hz. The question then becomes, is this match of the predominant human Alpha wave with the planetary Schumann resonances of electromagnetic energy coincidence, or is it part of biological evolutionary design? If the psyche has a spectral frequency component, then it may be possible that there have evolved in the human biosystem physiological modulators of this frequency component.

Microwave cavities are to the invisible frequency energy spectrum what lenses are to visible light energy.[416] In a biomedical engineering class lecture discussing the electrophysiology of the nervous system, I noticed the striking similarities between the waveguide horn antennas designed in another class, a microwave communication lab, and the images of the ventricular cavities within the human cranial cavity. The professor (a medical doctor and PhD) dismissed this possibility, stating that the ventricular system is seen by surgeons as having but two functions: (1.) it exists to help maintain thermal equilibrium in the brain, and (2.) it acts as mechanical protection for the brain, by attenuating sudden shock trauma to the skull. Regardless of these assumptions, the ventricular cavities (Fig. 29 & 30) can be seen to be dimensionally consistent with the microwave band of electromagnetic radiation, which covers a range of frequencies in wavelengths ranging from millimeters to centimeters.[417]

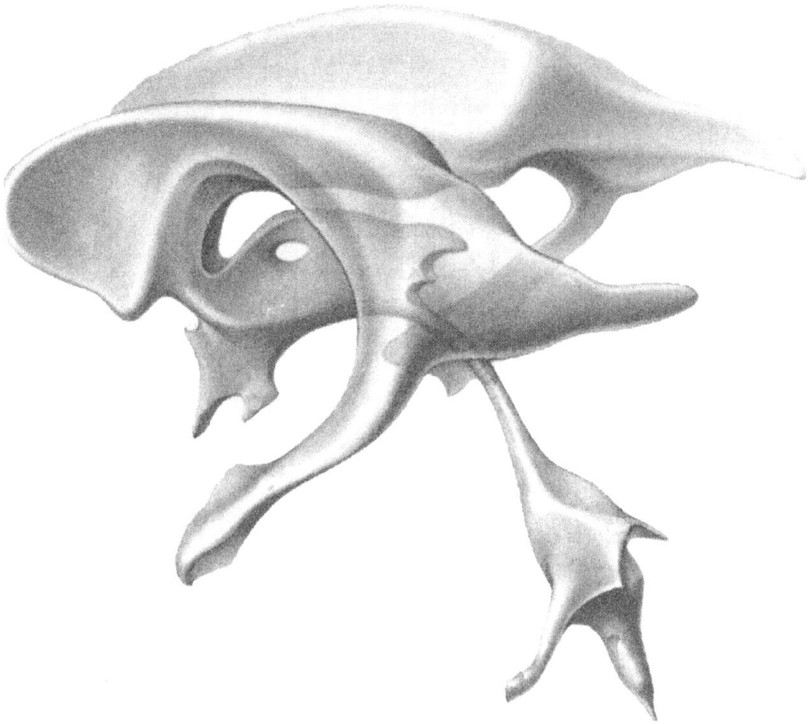

Figure 29. *Drawing of ventricular cavities. View from upper left rear of human cranium.*[418]

Another clue may be seen in cerebrospinal fluid itself, conventionally dismissed as being understood simply to act as a shock absorber or nutrient circulation system for the brain:

> Cerebrospinal fluid (CSF) is a clear, colorless liquid that bathes the brain and spinal cord. While the primary function of CSF is to cushion the brain within the skull and serve as a shock absorber for the central nervous system, CSF also circulates nutrients and chemicals filtered from the blood and removes waste products from the brain.[419]

Has nature evolved these amazingly shaped cavities simply to cushion the brain from a blow to the head? And why do they appear so similar, in shape and size, to conventional microwave horn antennas?

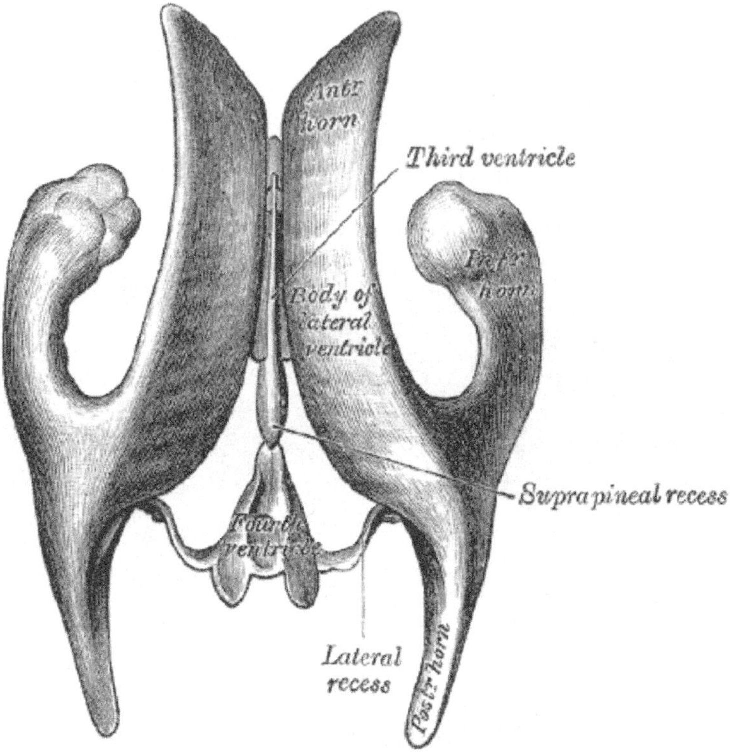

Figure 30. *Drawing of ventricular cavities. View from above human cranium.*[420]

Comparing the drawing of the ventricular cavities (Fig. 30) with the image of a waveguide system for microwaves used in radar (see Fig. 31, next page), it can be surmised that perhaps within the cavities of the ventricular system at the heart of the human brain system can be discovered another product of millions of years of an evolving human sensory system, perhaps a projectorium, a space where the mind builds, creates, designs, and visualizes, that has not yet been explored by science due to "looking in the wrong places."

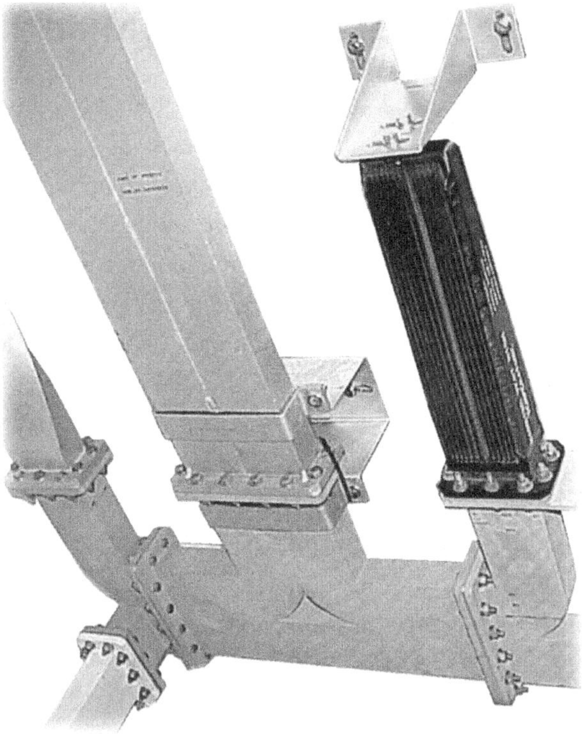

Figure 31. *Microwave radar circuitry.*[421]

Chapter 5: Tuning the Mind

This book has presented a theory that the ground of consciousness, indeed, consciousness itself, manifests as a continuously-transforming flow of energy in a two-way inward/outward flow between two domains whose boundaries are omni-present. This energy flux is encoded with both information and meaning, flowing alternately between the explicate domain of space–time and the nonspatial, nontemporal transcendent domain which Bohm termed the implicate order, a domain that exists within the center, everywhere, of space–time. In order to express the interconnectedness of this energy flux, spanning a wholeness that embraces both space-time phenomena and also nondual transcendent phenomena, Pribram selected the term *holoflux*, from the root term holism, which has been defined as:

> the fundamental interconnectedness of all phenomena, and within this fundamental interconnectedness not only is each entity, relationship, experience and phenomenon a hollow or whole, in its own right, but all wholes are held together by a great unifying force.[422]

Bohm himself referred to this as "the Whole" or "Wholeness."[423]

The Holoflux Theory of Consciousness

The *holoflux theory of consciousness* rests upon two underlying developments in modern science: Karl Pribram's holonomic brain theory, and David Bohm's ontological interpretation of quantum theory. Their theories, discussed in this book, have been found to be congruent in an approach to mapping and understanding consciousness. They are also supported by, and do not violate, established principles of electrical engineering. The holoflux theory proposed here can be categorized as a cosmological ontological process structured as follows:

- A CENTERED COSMOS: In agreement with the theory of general relativity, the holoflux theory is founded upon the notion that the universe has a center, and that this center is in actuality, present everywhere.

- HOLOFLUX: The dark energy of the implicate order, outside of space–time. Pribram's theorized flux within Bohm's implicate order is an energy of sentient awareness that views the outwardly explicate world of space–time from within an implicate order.

- QUANTUM BLACK HOLES: Also known as *Planck holospheres* of the Bohmian implicate order; bounded by spherical shells of the Planck length diameter 10^{-35} meters.

- THE HOLOPLENUM: An invariant *plenum* of close-packed holospheres underlying, and from which is projected, the explicate order. The holoplenum fills and encompasses continuum of isospheres everywhere in the physical universe from the center of each position in space–time. This concept was intuited by Leibniz in his theory of the monads.

- RESONANCE: Holoflux frequency resonance connects the explicate order (the outside world), with the implicate order (the inside world). Holoflux communication occurs everywhere *between* the implicate order (frequency-phase flux in the implicate order, f_i) and the resonating isospheres of the explicate order, manifesting in spherical shells. These shells are separated at fixed corresponding distances, λ_i, from the center of each Planck holosphere, thus maintaining the relationship, $f_i \lambda_i$,= c, where c is the speed of light in a vacuum (~3×10^8 m/sec).

- CYCLIC PROCESS: A flowing *electromagnetic-holoflux* movement of energy continually cycles between the implicate order and the explicate order; the flow can be mathematically modeled through Fourier transforms and the Bohm quantum potential wavefunctions.

- A COSMIC CLOCK: Granular units of Planck time (t_P), oscillate at a "clock-cycle" of 5.39106 x 10^{44} cycles per second, synchronizing the flow between alternating manifestations of implicate and explicate frequency information. In Cartesian coordinates, this can be pictured as a square wave, a Planck-time clock.

- BOHM'S QUANTUM POTENTIAL FUNCTION: these are conceptualized computations within the implicate order, and are "expressed" in space–time (the explicate order) via the effect of Bohm's quantum potential, *Q*.

- A FOURIER LENS: A Fourier-type mathematical transform can be seen to act as the lens through which the implicate "views" or "cognizes" the explicate order of space–time. Fourier lensing is the process by which the implicate order retrieves, observes, and experiences information generated within the explicate order.

A diagram of these assumptions is presented in Fig. 32, "Solving the Hard Problem of Consciousness," in which the relative positions of essential concepts from the following five philosophers have also been indicated:

- David Chalmers' "structure of awareness" is posited to be synonymous with the implicate order, and his "structure of consciousness" accordingly is shown within the explicate order.[424]

- Ned Block's "A-consciousness" lies in the implicate order, and his "B-consciousness" in the explicate order.[425]

- David Bohm's "meaning" lies in the implicate order, and his "information" in the explicate order.[426]

- Pierre Teilhard de Chardin's "Omega" lies in the implicate order, and his "centro-complexity" in the explicate order.[427]

- Ervin László's "A-dimension" lies in the implicate order, and his "B-dimension" lies in the explicate order.[428]

In each of these examples, the process can be visualized as shell-like waves of radiant information-embedded energy spreading out and intersecting with the myriad of the other expanding shells of active information from all of the other "particle/holospheres" in the holoplenum-filled cosmos. The intersection of these waves and their condensation where the density of intersection is maximum, is regarded in classical physics as "matter," but this can also be seen as a process, an on-going holographic extrusion of information *from* the implicate order *into* the explicate order, where the manifestations in space–time are holographically projected.

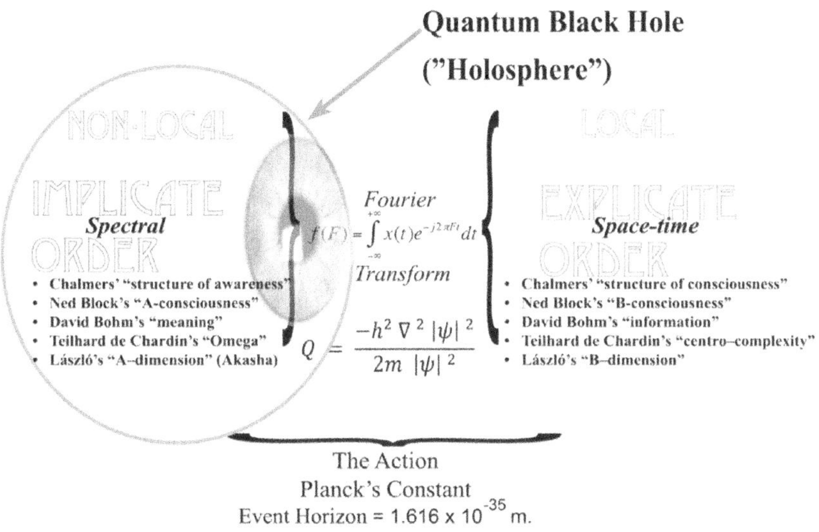

Figure 32. *Solving the hard problem of consciousness.*

In the diagram, an iris-like lens at the Planck length aperture, between Bohm's implicate and explicate order, mirrors Pribram's 1991 conceptualization in *Brain and Perception*:

> These two domains characterize the input to and output from a lens that performs a Fourier transform. On one side of the transform lies the space–time order we ordinarily perceive. On the other side lies a distributed enfolded holographic-like order referred to as the frequency or spectral domain.[429]

Note that the image of an iris in (Fig. 32) appears at the edge of the event horizon of a quantum black hole, or implicate order holosphere. The iris symbolizes consciousness looking out from within the implicate order into space–time via a Fourier transform lensing process. This can also be understood as the "observer observing the observed," as in the context of Wheeler's "Self-observing U."[430] Two equations can be seen in the center of the figure: Pribram's holonomic Fourier transform, and Bohm's quantum potential function *Q*. These illustrate the ongoing mathematical lensing–flow of conscious energy between the local (explicate) and nonlocal (implicate) regions of reality.

In projecting the cosmos, nonlocal holoflux energy in the spectral domain resonates actively with electromagnetic plasma in the space–time domain; thus the nonlocal implicate order affects the local, explicate order. From its source in the nonlocal implicate order, information emerges into space–time as electromagnetic energy at the Planck wavelength. This information–energy travels outwardly from the implicate centers at the holoplenum level, moving as spherical shells, waves of electromagnetic energy continuously manifesting form as they intersect holographically in our local, space–time, explicate regions.

The flow of holoflux energy into space–time can also be expressed visually, using classical electrical engineering imagery, as a variation of Fleming's Left Hand Rule (Fig. 33).[431]

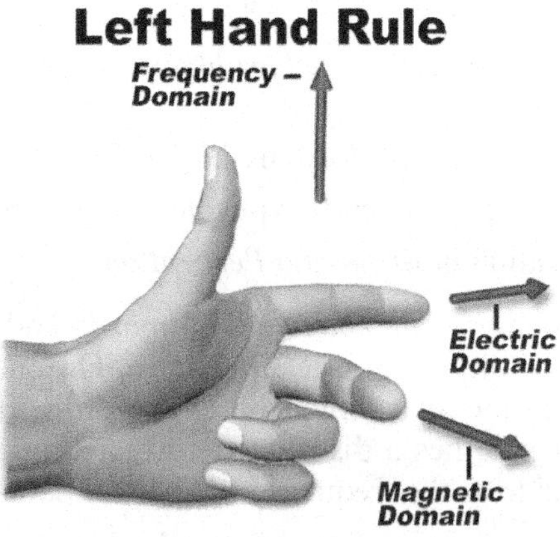

Figure 33. *Left–Hand Rule.*

The three domains can be visualized with the thumb, index finger, and middle finger of the left hand held mutually perpendicular to indicate the three orthogonal fields: the electric domain (e_d), the magnetic domain (m_d), and the frequency flux domain (f_d). We hypothesize that all three domains have equally valid ontological reality, but that only two of them, the magnetic and the electric, manifest explicitly in the space–time domain. Conversely, the third domain, the frequency flux domain (f_d), manifests "within" the implicate domain, transcendent of space and time.

The current dominant research paradigm focuses upon the tangential electrical field domain (e_d), operating in the system that powers the so-called "meat-mind" of neuronal fibers in brain and nervous systems during waking cognitive consciousness. The other mode of consciousness, operating at right angles to the electric, is the magnetic, radial or axial energy in various degrees of resonant connection with the implicate center. Thus two systems of consciousness are seen to coexist simultaneously, complementary, at right angles to one another.

The holoflux model identifies consciousness in the magnetic domain (m_d) as having a locus throughout the cardiovascular system. The entire system of magnetically polarized molecules in the bloodstream, throughout the more than 25,000 miles of veins and capillaries provides the infrastructure for a fully connected plasma waveguide system at infrared light frequencies.[432] A parallel can be seen in twenty-first-century internet technology, where glass fiber optic cables interconnected over hundreds of thousands of miles act as waveguides for networks at visible light frequencies. Thus a magnetic infrared plasma network might currently be operational within the human body, providing different information processing capabilities and functions than those currently observed in the network of neurons in the brain.

Testing the Holoflux Theory

If consciousness is truly manifesting everywhere as an energy flow in electromagnetic-frequency fields, then one should be able to detect experimentally the frequency-band location of various distinct high bandwidth network channels within human physiology. In contrast, the widespread assumption of material science has been that consciousness is *only* associated with neuronal brain activity, and that it emerges from neuronal activity in the human brain only as an *epiphenomenon*.[433]

For testing the holoflux hypothesis, dimensional evidence indicates that candidate ranges for testing the electromagnetic field component of consciousness can be found within the infrared spectrum. This region lies just below the threshold of the visible spectrum that is picked up by human eye cone structures.[434] Human core body temperatures, ranging from 36.3 C to 37.5 C on a diurnal cycle, indicate that, applying Wien's Law, we should search the infrared spectrum in a bandwidth between 9.36425 microns and 9.32808 microns, near the middle of the infrared.

The range of human body temperature associated with human consciousness (indeed, human life), spans an exceptionally wide frequency band. Fig. 34 displays this range as spanning 9.88 to 9.11 microns, charted as corresponding wavelength. For reference, an image of a 10 micron diameter single red blood cell is shown to the left in the Figure.

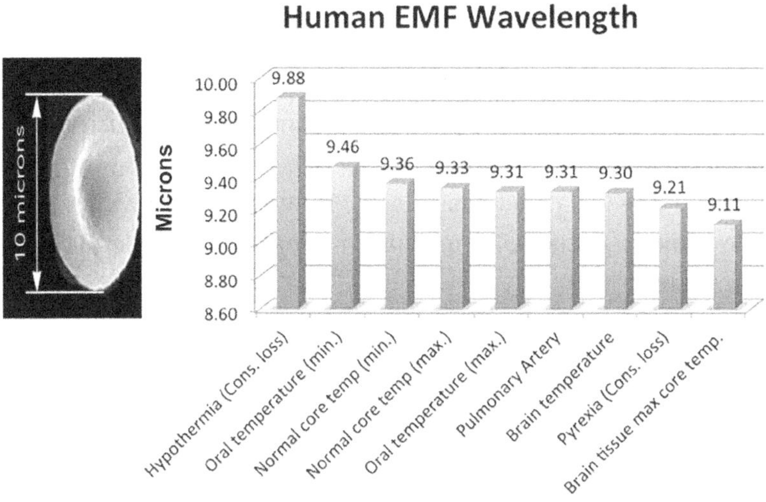

Figure 34. *Human electromagnetic wavelength.*[435]

Human body temperature ranges from a minimum of 68.0 F. (hypothermia with loss of consciousness) to a maximum of 106.7 F. (the maximum core brain temperature immediately before the onset of physiological death). Converting from wavelength, this range is equivalent to a frequency range of 30.3 gigahertz to 32.0 gigahertz, an enormously wide band compared to, for example, the FM radio broadcast frequency band, ranging from 87.0 to 107.1 MHz. Over 1000 equivalent FM radio bands, each filled with separate broadcast stations, could easily fit within the human infrared radiation band, with no overlapping interference.[436]

Assuming that electromagnetic energy in the infrared range networks consciousness within the human body, where might we search to detect such activity? One approach in the search for an infrared component of consciousness would be to monitor the dynamics of an infrared spectrum emanating from and external to the human body in an attempt to detect information laden photons escaping the body as modulated infrared radiation.

> Interestingly, because biologic materials are transparent to light in the near-infrared region of the light spectrum, transmission of photons through organs is possible.[437]

Another approach would be to search for infrared energy signals flowing as patterns within the various physiological waveguide channels within the human body. The ubiquitous blood capillary system, for example, with typical inner diameters of 10 microns, is a likely candidate to be acting as an electromagnetic waveguide. Capillaries provide a ready-made network infrastructure within which the flow and resonance of a modulated infrared energy plasma might be discovered.

Bandwidths other than the infrared, however, cannot not be ignored in the search for an electromagnetic energy mode of conscious activity. If we consider the much smaller microtubules which are embedded within the neurons, we find them to be excellent candidates to act as waveguides for electromagnetic frequency propagation at an even higher frequency band. In July 2015, Hameroff provided the following update on his research in the area of microtubules and anesthesiology, and new support for Pribram's holonomic brain/mind theory:

> Actually, the brain is looking more like an orchestra, a multi-scalar vibrational resonance system, than a computer. Brain information patterns repeat over spatiotemporal scales in fractal-like, nested hierarchies of neuronal networks, with resonances and interference beats. . . .Holograms encode distributed information as multi-scalar interference of coherent vibrations, e.g. from lasers. Pribram lacked a proper coherent source, a laser in the brain, but evidence now points to structures inside brain neurons called microtubules as sources of laser-like coherence for the brain's vibrational hierarchy? . . .microtubules have quantum resonances in gigahertz, megahertz and kilohertz frequency ranges (the work of Anirban Bandyopadhyay and colleagues at National Institute of Material Science in Tsukuba, Japan). . . .These coherent "fractal frequencies" in microtubules apparently couple to even faster, smaller-scale terahertz vibrations among intra-tubulin "pi electron resonance clouds," and to slower ones, e.g. by interference "beats" giving rise to larger scale EEG.[438]

Applying Wien's Law to the inner diameter of a typical microtubule results in a wavelength in the ultraviolet band, significantly higher in frequency than the 9–10 micron infrared, yet just on the other side of the "visible spectrum" band used by eye rods and cones during normal daylight visual reception. In 2010, a team exploring microtubule resonance at the University of Pavia's Nano-Biotechnology laboratory verified the existence of electromagnetic resonance of microtubules at a frequency of approximately 1.51 GHz.[439] The team concluded the following:

> The analysis of the results of bi-refringence experiment highlights that the microtubules react to electromagnetic fields . . . at a frequency of 1510 MHz . . . . Considering the nanoscopic size of microtubules, the resonance analysis would be more effective if carried out on much higher frequencies (up to 100 GHz), with suitable instrumentation. The presence of a small but sharp resonance effect at a low frequency could be the hint of a much more evident effect at even higher frequencies.[440]

Confirmation of the generation of such electromagnetic frequencies is described in a similar study completed in the following year at another research laboratory as reported in the *Journal of Theoretical Biology*, which describes additional evidence for high frequency electromagnetic field oscillations associated with microtubules:

> The electromagnetic radiation and field characteristics of the whole cellular microtubule network have not been theoretically analyzed previously. The results indicate that macroscopic detection system (antenna) are not suitable for measurement of cellular electrodynamic activity in the radiofrequency region since the radiation of single cells is very low (lower than 10^{-20} Watts). Low noise nanoscopic detection methods with high spatial resolution, enabling measurements in the cell vicinity, would be desirable, in order to measure cellular electrodynamic activity reliably.[441]

Some interest has already been shown in the function of microtubules as antennas, for example in ciliogenesis. Cilia are present on almost every cell type in the human body and are composed of a microtubule-based core structure, that have been seen to "act as antennae" capable of sensing extracellular signals, such as hormones, pheromones, odorants, and changes in the local environment.[442]

Several potential ranges for electromagnetic networks of consciousness are indicated in Fig. 35, "Eight Feasible Bands for Consciousness." Note that while microtubules are 40 nm in diameter, even smaller is the average 20 nanometers synaptic cleft width, which would make the interstitial space between synapses resonant in the ultraviolet range.

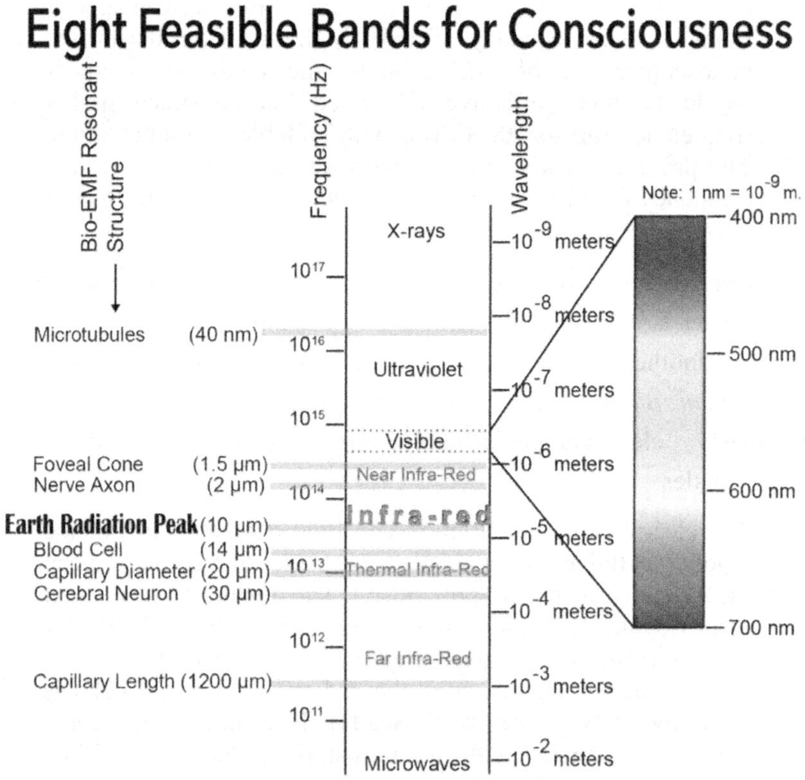

Figure 35. *Eight feasible bands for consciousness.*[443]

The adult human brain is estimated to contain "approximately 20 billion neocortical neurons, with an average of 7,000 synaptic connections each," making these structures possible candidates for involvement in processing and memory storage; every cubic centimeter of cerebral cortex contains roughly a trillion synapses.[444]

To test whether consciousness modulates electromagnetic field activity, we should begin with the monitoring, recording, and frequency domain analysis of electromagnetic spectra radiating from biological structures. We should also measure, record, and analyze what has previously been regarded only as signal "noise" in the infrared energy radiating through the blood capillary system.

Further analysis would consist of searching for modulated patterns encoded within these spectral recordings to identify any patterns that might indicate that they carry encoded information. This would be similar to the research methodology conducted by John Lilly in the 1950s, in a research project funded by the National Science Foundation. During this research, Lilly's team monitored and recorded audible dolphin/human communications, and applied Fourier frequency domain analysis in the search for supra-audible information patterns.[445]

In Conclusion: Uploading Consciousness

It is possible to speculate on the many practical implications of this holoflux theory for future technologies. Identification of an electromagnetic energy field manifesting information characteristics in biological systems would support the feasibility that, in principle, major components of consciousness might eventually be stored, maintained, and manipulated using hardware (e.g., fiber optics and silicon) instead of wetware (biological tissue). An intriguing implication would be that consciousness might be uploaded from wetware into such a hardware environment, and vice versa.

On the other hand, human wetware consciousness may even now be in the process of being uploaded continually into a planetary geomagnetic noosphere, tuned to human species-specific wavelengths, on a continual basis, fed by over 7 billion beating human hearts. This would be the source of a radiant holoflux energy feeding into an evolving planetary noosphere, part of the ever higher centro-complexities predicted by the French paleontologist Teilhard de Chardin in his theories of the noosphere and hyperphysics, articulated in his masterpiece, *The Human Phenomenon.* This will be discussed more fully in the second book in this series, *Tuning the Mind: Joining the Noosphere,* in which are developed pragmatic applications of the holoflux model of consciousness based upon the work of Rudolf Steiner, Teilhard de Chardin, Sri Aurobindo, John Lilly and other psychonauts of the twentieth century.[446]

About the Author

Shelli Renée Joye was born on the island of Trinidad in the British West Indies, and grew up in London and northern Virginia. Accepting a scholarship in Physics from Rice University, she completed a Bachelor of Science in Electrical Engineering. After working with Dr. John Lilly in research involving interspecies communication between dolphins and humans, she began a lifelong exploration of the interface between consciousness, yoga, electromagnetic theory, and mathematics.

She was awarded an M.A. in Asian Philosophy from the California Institute of Asian Studies under the guidance of the founder, Dr. Haridas Chaudhuri, and published a Master's thesis, "The History, Philosophy, and Practice of Tantra in South India." In the early '90s she studied with Fr. Bede Griffiths at Shantivanam in South India.

Dr. Joye completed a Ph.D. program in the Philosophy, Cosmology, and Consciousness program at the California Institute of Integral Studies under the guidance of Drs. Allan Leslie Combs, Brian Swimme, and Dean Radin. She currently lives in a cedar cabin in the Cascade mountains near Lassen forest in northeastern California.

About the Viola Institute

The Viola Institute is a non-profit corporation established to promote consciousness research and to foster an integral, ecumenical, science-based approach to spiritual traditions, religious practices, and the numinous arts. To this end, the Institute sponsors art exhibitions and publishes works of nonfiction around the themes of contemplative practice, comparative religion, and consciousness studies. A primary goal of the Viola Institute is to develop a consciousness research and retreat center on 33 acres of land in Gualdo di Sopra, a forested countryside near Assisi, Italy. For more information, or to make a tax deductible contribution contact:

The Viola Institute
35366 Wild Turkey Lane
Viola, CA 96088

REFERENCES

Alabiso, Carlo, and Ittay Weiss. 2015. *A Primer on Hilbert Space Theory: Linear Spaces, Topological Spaces, Metric Spaces, Normed Spaces, and Topological Groups*. New York: Springer.

Alkari. 2012. "File:Yin_yang.svg" (graphic file). November 28, 2015. Wikimedia Commons. Retrieved from https://commons.wikimedia.org/wiki/File:Yin_yang.svg.

Allaby, Michael, and Ailisa Allaby, eds. 1999. "Telluric Current." In *A Dictionary of Earth Sciences*. 2nd ed. New York: Oxford University.

Aur, Dorian, and Mandar S. Jog. 2010. *Neuroelectrodynamics: Understanding the Brain Language*. Amsterdam: IOS Press.

Aurobindo Ghose. 1997. *Tales of Prison Life*. Pondicherry, India: Sri Aurobindo Ashram.

Averse. 2007. "File:Diplexer1.jpg" (graphic file). November 13, 2015. Wikimedia Commons. Retrieved from https://en.wikipedia.org/wiki/Microwave#/media/File:Diplexer1.jpg.

Baggott, Jim. 2011. *The Quantum Story: A History in 40 Moments*. New York: Oxford University Press.

Bain, George Grantham. 2011. "File:Niels_Bohr_Date_Unverified_LOC.jpg" (graphic file). November 28, 2015. Wikimedia Commons. Retrieved from https://commons.wikimedia.org/wiki/File:Niels_Bohr_Date_Unverified_LOC.jpg.

Bekenstein, Jacob D. 1973. "Black Holes and Entropy." *Physical Review* 7 (8): 2333–46.

Bell, E. T. 1937. *Men of Mathematics*. New York: Simon and Schuster.

Bell, J. S. 1987. *The Speakable and Unspeakable in Quantum Mechanics*. Cambridge: Cambridge University Press.

Benenson, Walter, John W. Harris, Horst Stocker, and Lutz Holger, eds. 2006. *Handbook of Physics*. New York: Springer Science.

Bergson, Henri. 1911. *Creative Evolution*. Translated by Arthur Mitchell. New York: Henry Holt.

Bisson, Terry. 1990. "They're Made Out of Meat." *Omni Magazine* 13 (7): 42–45. New York: General Media.

Block, Ned. 2007. *Consciousness, Function, and Representation: Collected Papers, Vol. 1*. Cambridge: The MIT Press.

Bohm, David. 1951. *Quantum Theory*. New York: Prentiss-Hall.

———. 1952. "A Suggested Interpretation of the Quantum Theory in Terms of 'Hidden Variables,' Vol. 1." *Physical Review* 85 (2): 166–93. Retrieved from http://fma.if.usp.br/~amsilva/Artigos/p166_1.pdf.

———. 1965. *The Special Theory of Relativity*. Philadelphia: John Benjamins.

———. 1978. "The Enfolding-Unfolding Universe: A Conversation with David Bohm." In *The Holographic Paradigm and Other Paradoxes: Exploring the Leading Edge of Science*, edited by Ken Wilber, 44–104. Boulder, CO: Shambhala.

———. 1979. *A Question of Physics: Conversations in Physics and Biology*, edited by P. Buckley and F. David Peat. London: Kegan Paul.

———. 1980. *Wholeness and the Implicate Order*. London: Routledge.

———. 1985. *Unfolding Meaning: A Weekend of Dialogue with David Bohm*. London: Routledge.

———. 1986. "The Implicate Order and the Super-Implicate Order." In *Dialogues with Scientists and Sages: The Search for Unity*, edited by Renée Weber, 23–49. New York: Routledge.

———. 1987. "Hidden Variables and the Implicate Order." In *Quantum Implications: Essays in Honour of David Bohm*, edited by B. J. Hiley and F. David Peat, 33–45. London: Routledge.

———. 1989. "Meaning and Information." In *The Search for Meaning: The New Spirit in Science and Philosophy*, edited by Paavo Pylkkänen, 43–85. Northamptonshire: The Aquarian Press.

———. 1990a. "Beyond Limits: A Full Conversation with David Bohm." Interview by Bill Angelos for Dutch public television. Posted March 5, 2011. Retrieved from http://bohmkrishnamurti.com/beyond-limits/

———. 1990b. "A New Theory of the Relationship of Mind and Matter." *Philosophical Psychology* 3 (2): 271–86.

Bohm, David, and Basil J. Hiley. 1993. *The Undivided Universe: An Ontological Interpretation of Quantum Theory.* London: Routledge.

Bohm, David, and J. Krishnamurti. 1985. *The Ending of Time: Where Philosophy and Physics Meet.* New York: Harper Collins.

Bohm, David, and J. Krishnamurti. 1999. *The Limits of Thought: Discussions between J. Krishnamurti and David Bohm.* London: Routledge.

Bohm, David, and F. David Peat. 1987. *Science, Order, and Creativity.* London: Routledge.

Bohm, David, and R. Weber. 1982. "Nature as Creativity." *ReVision* 5 (2): 35–40.

Booth, J. C., S. A. Koren, and Michael A. Persinger. 2005. "Increased Feelings of the Sensed Presence and Increased Geomagnetic Activity at the Time of the Experience During Exposures to Transcerebral Weak Complex Magnetic Fields." *International Journal of Neuroscience* 115 (7): 1039–65.

Born, Irene, trans. 1971. *The Born–Einstein Letters: Friendship, Politics and Physics in Uncertain Times.* New York: Macmillan.

Borsellino, A., and T. Poggio. 1972. "Holographic Aspects of Temporal Memory and Optomotor Responses." *Kybernetik* 10 (1): 58–60.

Broughton, S. Allen, and Kurt M. Bryan. 2008. *Discrete Fourier Analysis and Wavelets: Applications to Signal and Image Processing.* New York: Wiley.

Browder, Andrew. 1996. *Mathematical Analysis: An Introduction.* New York: Springer-Verlag.

Bruskiewich, Patrick. 2014. *Max Planck and Black-Body Radiation.* Vancouver: Pythagoras Publishing.

Carr, Bernard J. and Steven B. Giddings. 2005. "Quantum Black Holes." *Scientific American* 292 (5): 30–35.

Cazenave, Michel, ed. 1984. *Science and Consciousness: Two Views of the Universe, Edited Proceedings of the France-Culture and Radio-France Colloquium, Cordoba, Spain*. Oxford: Pergamon.

Chalmers, David J. 1995. "Facing Up to the Problem of Consciousness." *Journal of Consciousness Studies* 2 (3): 200–19.

———. 2010. *The Character of Consciousness*. New York: Oxford University.

Chaudhuri, Haridas. 1954. *The Philosophy of Integralism: The Metaphysical Synthesis in Sri Aurobindo's Teaching.* Pondicherry, India: Sri Aurobindo Ashram Trust.

———. 1969. *The Integral Philosophy of Sri Aurobindo*. London: George Allen & Unwin.

Chen, Frances F. 2006. *Introduction to Plasma Physics and Controlled Fusion: Vol 1. Plasma Physics*. 2nd ed. New York: Springer.

Cheney, Margaret. 1981. *Tesla: Man Out of Time*. New York: Dorset Press.

Clark, Walter. 1939. *Photography by Infrared—Its Principles and Applications.* New York: John Wiley & Sons.

Cohn, Stephen M., 2007. "Near-Infrared Spectroscopy: Potential Clinical Benefits in Surgery." *Journal of the American College of Surgeons* 205 (2): 322–32. doi: 10.1016/j.jamcollsurg.2007.02.024.

Collister, Rupert Clive. 2010. *A Journey in Search of Wholeness and Meaning*. New York: Peter Lang.

Cook, David M. 2002. *The Theory of the Electromagnetic Field.* Englewood Cliffs, NJ: Prentice-Hall.

Crease, Robert P. 2008. *The Great Equations: Breakthroughs in Science from Pythagoras to Heisenberg*. New York: W. W. Norton.

Darekk2. 2012. "File:Water_infrared_absorption_coefficient_large.gif" (graphic file). November 28, 2015. Wikimedia Commons. Retrieved from https://en.wikipedia.org/wiki/File:Water_infrared_absorption_coefficient_large.gif.

Darling, David J. 2004. *The Universal Book of Mathematics: From Abracadabra to Zeno's Paradoxes*. Hoboken, NJ: Wiley.

Dawson, Lorne L. 2006. *Comprehending Cults: The Sociology of New Religious Movements*. New York: Oxford University.

Dennett, Daniel C. 1991. *Consciousness Explained*. New York: Back Bay Books.

Dewey, B. 1985. *The Theory of Laminated Spacetime*. Inverness, CA: Bartholomew.

Dewynne, Dustin. 2012. "File:Dualism-vs-Monism.png" (graphic file). November 27, 2015. Wikimedia Commons. Retrieved from https://commons.wikimedia.org/wiki/File:Dualism-vs-Monism.png.

Dorf, Richard C., ed. 1997. *The Electrical Engineering Handbook*. 2nd ed. Boca Raton, FL: CRC Press.

Drachman, David A. 2005. "Do We Have Brain to Spare?" *Neurology* 64 (12): 2004–5. doi: 10.1212/01.WNL.0000166914.38327.BB.

Edelman, Gerald M., and Giulio Tononi. 2000. *A Universe of Consciousness: How Matter Becomes Imagination*. New York: Basic Books.

Edmondson, Amy C. 1987. *A Fuller Explanation: The Synergetic Geometry of R. Buckminster Fuller*. Boston: Birkhäuser.

Eliot, Thomas Stearns. 1943. *Four Quartets*. New York: Harcourt Brace.

———. 1976. *Collected Poems 1909–1962*. New York: Harcourt Brace.

Ferrer, Jorge, and Jacob Sherman, eds. 2008. *The Participatory Turn: Spirituality, Mysticism, Religious Studies.* Albany: State University of New York.

Feuerstein, Georg. 1987. *Structures of Consciousness: The Genius of Jean Gebser—An Introduction and Critique*. Lower Lake, CA: Integral.

Fields, R. Douglas. 2009. *The Other Brain*. New York: Simon & Schuster.

Fourier, Jean Baptiste Joseph. 1822. *The Analytic Theory of Heat*. Paris: Firmin Didot Père et Fils.

Fulvio314. 2013. "File:Coat_of_Arms_of_Niels_Bohr.svg" (graphic file). November 28, 2015. Wikimedia Commons. Retrieved from https://commons.wikimedia.org/wiki/File:Coat_of_Arms_of_Niels_Bohr.svg.

Gabor, Dennis. 1946. "Theory of Communication." *Journal of the Institute of Electrical Engineers* 93: 429–41.

Garay, Luis J. 1995. "Quantum Gravity and Minimum Length." *International Journal of Modern Physics* 10 (2): 145–65.

Gebser, Jean. 1949. *The Ever-Present Origin: Part One: Foundations of the Aperspectival World.* Translated by J. Keckeis. Stuttgart, Germany: Deutsche Verlags-Anstalt.

———. 1956 (1996). "Cultural Philosophy as Method and Venture." Translated by Georg Feuerstein. *Integrative Explorations: Journal of Culture and Consciousness* 3 (1): 77–82. Retrieved from http://static1.squarespace.com/static/535ef5d8e4b0ab57db4a06c7/t/541f74a0e4b0394ddbf5a040/1411347616917/integrative_explorations_3.pdf.

Geiger, Rudolf, Robert H. Aron, and Paul Todhunter. 2003. *The Climate Near the Ground.* Lanham, MD: Rowman and Littlefield.

Gidley, Jennifer. 2007. "The Evolution of Consciousness as a Planetary Imperative: An Integration of Integral Views." *Integral Review* 3 (5): 4–226. Retrieved from http://integral-review.org/pdf-template-issue.php?pdfName=issue_5_gidley_the_evolution_of_consciousness_as_a_planetary_imperative.pdf.

Glatzmaier. 2007. "File:Geodynamo_Between_Reversals.gif" (graphic file). November 12, 2015. Wikimedia Commons. Retrieved from https://commons.wikimedia.org/wiki/File:Geodynamo_Between_Reversals.gif.

Globus, Gordon G. 2006. *The Transparent Becoming of World: A Crossing Between Process Philosophy and Quantum Neurophilosophy*. Philadelphia: John Benjamins.

Gott, J. Richard III, Mario Jurić, David Schlegel, and Fiona Hoyle. 2005. "A Map of the Universe." *The Astrophysics Journal* 624 (2): 463–514.

Gray, Henry. 1918. "File:Gray754.png" (graphic file). October 31, 2015. Wikimedia Commons. Retrieved from https://commons.wikimedia.org/wiki/File:Gray754.png.

Haas, Jonathan. 2012. "File:Neurons_uni_bi_multi_pseudouni.svg" (graphic file). October 31, 2015. Wikimedia Commons. Retrieved from http://commons.wikimedia.org/wiki/File:Neurons_uni_bi_multi_pseudouni.svg.

Hameroff, Stuart R. 2015. "Is Your Brain Really a Computer, or Is It a Quantum Orchestra?" *Huffington Post*, July 9. Retrieved from http://www.huffingtonpost.com/stuart-hameroff/is-your-brain-really-a-co_b_7756700.html.

Hameroff, Stuart, and Roger Penrose. 1996. "Conscious Events as Orchestrated Space-Time Selections." *Journal of Consciousness Studies* 3 (1): 35–53.

Hameroff, Stuart R., Travis J. A. Craddock, and Jack A. Tuszynski. 2014. "Quantum Effects in the Understanding of Consciousness." *Journal of Integrative Neuroscience* (13) 2: 229–52. doi:10.1142/S0219635214400093.

Havelka, D., M. Cifra, O. Kucera, J. Pokorny, and J. Vrba. 2011. "High-frequency Electric Field and Radiation Characteristics of Cellular Microtubule Network." *Journal of Theoretical Biology* 286 (7): 31–40.

Heehs, Peter. 2008. *The Lives of Sri Aurobindo*. New York: Columbia University.

Herculano-Houzel, Suzana. 2009. "The Human Brain in Numbers: A Linearly Scaled-up Primate Brain." *Frontiers in Human Neuroscience* (3) 31: 1–11. doi:10.3389/neuro.09.031.2009.

Hertz, Heinrich. 1893. *Electric Waves: Being Researches On the Propagation of Electric Action with Finite Velocity Through Space*. London: MacMillan.

Hiley, B. J., and F. David Peat, eds. 1987. *Quantum Implications: Essays in Honour of David Bohm*. London: Routledge.

Hiley, B. J., and F. David Peat, eds. 1987. "The Development of David Bohm's Ideas from the Plasma to the Implicate Order." In *Quantum Implications: Essays Honour of David Bohm*, edited by B. J. Hiley and F. David Peat, 1–32. London: Routledge.

Howell, Kenneth B. 2001. *Principles of Fourier Analysis*. Boca Raton, FL: Chapman & Hall.

Ishikawa, Hiroaki, and Wallace F. Marshall. "Ciliogenesis: Building the Cell's Antenna." *Nature Reviews Molecular Biology* 12: 222-34. doi:10.1038/nrm3085

Jahoe. 2012. "File:Electromagnetic-Spectrum.png" (graphic file). November 28, 2015. Wikimedia Commons. https://commons.wikimedia.org/wiki/File:Electromagnetic-Spectrum.png.

Jbarta. 2014. "File:Max_Planck_1878.GIF" (graphic file). November 28, 2015. Wikimedia Commons. Retrieved from https://commons.wikimedia.org/wiki/File:Max_Planck_1878.GIF.

Jibu, Mari, and Kunio Yasue. 1995. *Quantum Brain Dynamics and Consciousness*. Philadelphia: John Benjamins.

———. 2003. "Quantum Brain Dynamics and Quantum Field Theory." In *Brain and Being: At the Boundary Between Science, Philosophy, Language and Arts*, edited by Gordon Globus, Karl Pribram, and Giuseppe Vitiello, 267–90. Philadelphia: John Benjamins.

Johnston, Sean F. 2006. *Holographic Visions: A History of New Science*. New York: Oxford University.

Joye, S. 2016. *The Pribram–Bohm Holoflux Theory of Consciousness: An Integral Interpretation of the Theories of Karl Pribram, David Bohm, and Pierre Teilhard de Chardin* (Doctoral dissertation). ProQuest Dissertations and Theses database. (UMI No. 1803306323)

Kafatos, Menas, Rudolph E. Tanzi, and Deepak Chopra. 2011. "How Consciousness Becomes the Physical Universe." *Journal of Cosmology* 14: 1318–1328. Retrieved from http://journalofcosmology.com/Consciousness140.html.

Kant, Emmanuel. 1933. *Critique of Pure Reason*. Edited and translated by Norman Kemp Smith. London: Macmillan.

Köhler, Wolfgang. 1969. *The Task of Gestalt Psychology*. New Jersey: Princeton University.

Köhler, Wolfgang, and Mary Henle. 1971. *The Selected Papers of Wolfgang Köhler*. New York: W. W. Norton.

Kuehn, Kerry. 2014. A Student's Guide Through the Great Physics Texts, Vol. 1: The Heavens and The Earth. New York: Springer.

Kuo, Franklin. 1962. *Network Analysis and Synthesis*. New York: John Wiley & Sons.

László, Ervin. 2014. *The Self-Actualizing Cosmos: The Akasha Revolution in Science and Human Consciousness*. Rochester, VT: Inner Traditions.

Ledrew, Glenn. 2001. "The Real Starry Sky," *Journal of the Royal Astronomical Society of Canada* (95): 25–68.

Leith, Emmet N., and J. Upatnieks. 1962. "Reconstructed Wavefronts and Communication Theory." *Journal of the Optical Society of America* 52 (10): 1123–30.

———. 1965. "Photography by Laser." *Scientific American* 212 (6): 24–35.

Lilly, John. 1967. *The Mind of the Dolphin: A Nonhuman Intelligence*. New York: Doubleday.

———. 1977. *The Deep Self: Consciousness Exploration in the Isolation Tank*. New York: Simon & Schuster.

Lundqvist, Stig, ed. 1992. *Nobel Lectures, Physics 1971–1980*. Singapore: World Scientific.

McCraty, Rollin. 2003. *The Energetic Heart: Bioelectromagnetic Interactions Within and Between People*. Boulder Creek, CA: Institute of HeartMath.

McFadden, Johnjoe. 2002a. "The Conscious Electromagnetic Information (CEMI) Field Theory: The Hard Problem Made Easy." *Journal of Consciousness Studies* 9 (8): 45–60.

———. 2002b. "Synchronous Firing and Its Influence on the Brain's Electromagnetic Field: Evidence for an Electromagnetic Field Theory of Consciousness." *Journal of Consciousness Studies* 9 (4): 23–50.

———. 2007. "Conscious Electromagnetic Field Theory." *NeuroQuantology* 5 (3): 262–70.

McIntosh, Steve. 2007. *Integral Consciousness and the Future of Evolution*. St. Paul, MN: Paragon House.

Merrell-Wolff, Franklin. 1973. *The Philosophy of Consciousness Without an Object: Reflections on the Nature of Transcendental Consciousness*. New York: Julian Press.

Milonni, Peter W., and Joseph E. Eberly. 2010. *Laser Physics*. Hoboken, NJ: John Wiley & Sons.

Morin, Edgar. 1999. *Seven Complex Lessons in Education for the Future*. Translated by Nidra Poller. Paris: UNESCO.

National Multiple Sclerosis Society. n.d. "Cerebrospinal Fluid (CSF)." Retrieved from http://www.nationalmssociety.org/Symptoms-Diagnosis/Diagnosing-Tools/Cerebrospinal-Fluid-(CSF).

Netter, F. H. 1972. *A Compilation of Paintings of the Normal and Pathologic Anatomy of the Nervous System*. Summit, NJ: CIBA.

Neville, Katherine. 1992. "Saral and David Bohm, Prague, June 1992" (photograph). Retrieved from http://www.karlpribram.com/photos/.

Nilson, Arthur R., and J. L. Hornung. 1943. *Practical Radio Communication: Principles, Systems, Equipment, Operation, Including Very High and Ultra High Frequencies and Frequency Modulation*. 2nd ed. New York: McGraw-Hill.

Nishikawa, K., and M. Wakatani. 2000. *Plasma Physics*. Berlin: Springer-Verlag.

Oates, B. 1971. *Celebrating the Dawn: Maharishi Mahesh Yogi and the TM Technique*. New York: Putnam Books.

Ouspensky, P. D. 1949. *In Search of the Miraculous: Fragments of an Unknown Teaching*. London: Harcourt.

Oyster, Clyde W. 1999. *The Human Eye: Structure and Function*. Sunderland, MA: Sinauer Associates.

Peat, F. David. 1997. *Infinite Potential: The Life and Times of David Bohm*. Reading, MA: Addison-Wesley.

Penfield, Wilder. 1975. *The Mystery of the Mind: A Critical Study of Consciousness and the Human Brain*. Princeton, NH: Princeton University.

Persinger, Michael A. 2014. "Schumann Resonance Frequencies Found Within Quantitative Electroencephalographic Activity: Implications for Earth-Brain Interactions." *International Letters of Chemistry, Physics and Astronomy* 30: 24–32. doi:10.18052/www.scipress.com/ILCPA.30.24.

Pizzi, Rita, Giuliano Strini, Silvia Fiorentini, Valeria Pappalardo, and Massimo Pregnolato. 2010. "Evidences of New Biophysical Properties of Microtubules." In *Artificial Neural Networks*, edited by Scoyun J. Kwon. New York: Nova Science Publishers, 1–17. Retrieved from https://air.unimi.it/retrieve/handle/2434/167480/168890/evidences.pdf.

Planck, Max. 1901. "On the Law of Distribution of Energy in the Normal Spectrum." *Annalen der Physik* 309 (3): 553–63.

Pockett, Susan. 2000. *The Nature of Consciousness: A Hypothesis.* Lincoln, NE: Writers Club.

———. 2007. "Difficulties with the Electromagnetic Field Theory of Consciousness: An Update." *NeuroQuantology* 5 (3): 271–75. Retrieved from http://www.neuroquantology.com/index.php/journal/article /view/136/136 (login required).

Pribram, Karl H. 1962. "The Neuropsychology of Sigmund Freud." In *Experimental Foundations of Clinical Psychology*, edited by Arthur J. Bachrach, 442–68. New York: Basic Books.

———. 1971. *Languages of the Brain: Experimental Paradoxes and Principles in Neuropsychology.* Englewood Cliffs, NJ: Prentice-Hall.

———. 1982. "What the Fuss Is All About." In *The Holographic Paradigm and Other Paradoxes*, edited by Ken Wilber, 27–34. Boulder: Shambhala.

———. 1984. "Mind, Brain and Consciousness: The Organization of Competence and Conduct." In *Science and Consciousness: Two Views of the Universe*, edited by Julian Davidson and Richard Davidson, 115–32. New York: Springer.

———. 1990. "Prolegomenon for a Holonomic Brain Theory." In *Synergetics of Cognition*, edited by H. Haken, 150–84. Berlin: Springer-Verlag.

———. 1991. *Brain and Perception: Holonomy and Structure in Figural Processing.* Hillsdale, NJ: Lawrence Erlbaum.

———. 1994. "What Is Mind that the Brain May Order It?" In *Proceedings of Symposia in Applied Mathematics: Proceedings of the Norbert Wiener Centenary Congress, Vol. 52*, edited by V. Mandrekar and P. R. Masani, 301–29. Providence, RI: American Mathematical Society.

———. 2004a. "Brain and Mathematics." In *Brain and Being: At the Boundary Between Science, Philosophy, Language and Arts*, edited by Gordon Globus, Karl Pribram, and Giuseppe Vitiello, 215–40. Philadelphia: John Benjamins.

———. 2004b. "Consciousness Reassessed." *Mind and Matter* 2 (1): 7–35.

———. 2011. "Karl Pribram: Bibliography." Retrieved from http://www.karlpribram.com/bibliography/.

———. 2013. *The Form Within: My Point of View*. Westport, CT: Prospecta Press.

Pribram, Karl H., A. Sharafat, and G. Beekman. 1984. "Frequency Encoding in Motor Systems." In *Human Motor Actions: Bernstein Reassessed*, edited by H. T. A. Whiting, 121–56. Amsterdam: Elsevier Science Publishers.

Pylkkänen, Paavo. 2007. *Mind, Matter, and the Implicate Order*. New York: Springer.

Ramon y Cajal, Santiago. 2007. "File: Purkinje_cell_by_Cajal.png" (graphic file). January 30, 2016. Wikimedia Commons. Retrieved from https://commons.wikimedia.org/wiki/File:Purkinje_cell_by_Cajal.png.

Rendek, Kimberly N., Raimund Fromme, Ingo Grotjohann, and Petra Fromme. 2013. "Self-Assembled Three-Dimensional DNA Nanostructure." *Acta Crystallographica Section F: Structural Biology and Crystallization Communications* 69 (2): 141–46. doi:10.1107/S1744309112052128.

Ringbauer, M., B. Duffus, C. Branciard, E. G. Cavalcanti, A. G. White, and A. Fedrizzi. 2015. "Measurements On the Reality of the Wavefunction." *Nature Physics* 11 (2): 249–254. doi:10.1038/nphys3233.

Romanes, G. J., ed. 1964. *Cunningham's Textbook of Anatomy*. 10th ed. New York: Oxford University.

Ruhenstroth-Bauer, Gerhard. 1993. "Influence of the Earth's Magnetic Field on Resting and Activated EEG Mapping in Normal Subjects." *International Journal of Neuroscience* 73 (3): 331–49.

Runehov, Anne, and Luis Oviedo, eds. 2013. *Encyclopedia of Sciences and Religions*. Dordrecht, Netherlands: Springer Netherlands.

Ruppert, L. 1956. *History of the International Electrotechnical Commission*. Geneva: Central Bureau of the IEC.

Sentman, Davis D. 1995. "Schumann Resonances." In *Handbook of Atmospheric Electrodynamics, Vol. 1*, edited by Hans Volland, 267–96. Boca Raton, FL: CRC Press.

Squire, Larry R., ed. 1998. *The History of Neuroscience in Autobiography, Vol. 2*. London: Academic Press.

The Stanford Daily. 1977. "Scientists to Speak." September 26, 8.

Stuart, C.I.J., Y. Takahashi, and H. Umezawa .1979. "Mixed System Brain Dynamics: Neural Memory as a Macroscopic Ordered State." *Foundations of Physics* (9): 301–7.

Susskind, Leonard. 2008. *The Black Hole War: My Battle with Stephen Hawking to Make the World Safe for Quantum Mechanics*. New York: Little, Brown and Company.

Svdmolen. 2005. "File:Red_White_Blood_cells.jpg" (graphic file). November 12, 2015. Wikimedia Commons. Retrieved from https://en.wikipedia.org/wiki/File:Red_White_Blood_cells.jpg.

Talbot, Michael. 1992. *The Holographic Universe*. New York: Harper Collins.

Teilhard de Chardin, Pierre. (1945) 1959. "Life and the Planets." In *The Future of Man*, translated by Norman Denny, 97–123. New York: Harper & Row.

———. (1954) 1956. "The Nature of the Point Omega." In *The Appearance of Man*, translated by J.M. Cohen, 271–73. New York: Harper & Row.

———. 2003. *The Human Phenomenon*. Translated and edited by Sarah Appleton-Weber. Portland, OR: Sussex Academic. First published 1955.

Tjlaxs. 2005. "File:YoungJamesClerkMaxwell.jpg" (graphic file). November 28, 2015. Wikimedia Commons. https://commons.wikimedia.org/wiki/File:YoungJamesClerkMaxwell.jpg.

Turgeon, Mary Louise. 2004. *Clinical Hematology: Theory and Procedures*. Baltimore, MD: Lippincott Williams & Wilkins.

Van Dokkum, Pieter G., and Charlie Conroy. 2010. "A Substantial Population of Low-Mass Stars in Luminous Elliptical Galaxies." *Nature* 468 (7326): 940–42. doi:10.1038/nature09578.

Vejvoda, Stanislav. 2003. *17th International Conference on Structural Mechanics in Reactor Technology*. Prague, Czech Republic: Czech Standard Institute.

Warren, Stephen G., and Richard E. Brandt. 2008. "Optical Constants of Ice from the Ultraviolet to the Microwave: A Revised Compilation." *Journal of Geophysical Research*, 113 (D14): 1047–57.

Wasson, Tyler, and Gert Brieger. 1987. *Nobel Prize Winners: A Biographical Dictionary*. New York: H. W. Wilson.

Weber, Renée. 1982. "The Physicist and the Mystic—Is a Dialogue Between Them Possible?" In *The Holographic Paradigm and Other Paradoxes: Exploring the Leading Edge of Science*, edited by Ken Wilber, 187–214. Boulder, CO: Shambhala.

Wheeler, John Archibald. 1990a. *Information, Physics, Quantum: The Search for Links*. Austin, TX: University of Texas.

———. 1990b. *A Journey into Gravity and Spacetime*. New York: Scientific American.

———. 1998. *Geons, Black Holes, and Quantum Foam: A Life in Physics*. New York: W. W. Norton.

Wick, Manfred, Wulf Pinggera, and Paul Lehmann. 2003. *Clinical Aspects and Laboratory Iron Metabolism*. Vienna, Austria: Springer-Verlag.

Wilber, Ken, ed. 1982. *The Holographic Paradigm and Other Paradoxes: Exploring the Leading Edge of Science*. Boulder, CO: Shambhala.

———, ed. 1985. *Quantum Questions: Mystical Writings of the World's Great Physicists*. Boston: Shambhala.

———. 1996. *A Brief History of Everything*. Boston: Shambhala.

———. 1997. *Integral Spirituality: A Startling New Role for Religion in the Modern and Postmodern World*. Boston: Integral Books.

Wiltschko, Wolfgang, and Roswitha Wiltschko. 2008. "Magnetic Orientation and Magnetoreception in Birds and Other Animals." *Journal of Comparative Physiology A: Neuroethology, Sensory, Neural, and Behavioral Physiology* 191 (8): 675–93. doi:10.1007/s00359-005-0627-7.

Wiseman, Howard M., Fuwa Maria, Shuntaro Takeda, Marcin Zwierz, and Akira Furusawa. 2015. "Experimental Proof of Nonlocal Wavefunction Collapse for A Single Particle Using Homodyne Measurements." *Nature Communications* 6: 192–203. Retrieved from http://www.nature.com/ncomms/2015/150324/ncomms7665/full/ncomms7665.html.

Wong, Eva. 1997. *The Shambhala Guide to Taoism*. Boston: Shambhala.

Yau, Shing-Tung, and Steve Nadis. 2010. *The Shape of Inner Space: String Theory and the Geometry of the Universe's Hidden Dimensions*. New York: Basic Books.

Zimmer C. 2009. "Origins: On The Origin of Eukaryotes." *Science* 325 (5941): 666–68.

Zizzi, Paola. 2006. "Consciousness and Logic in a Quantum Computing Universe." In *The Emerging Physics of Consciousness*, edited by Jack A. Tuszyński, 457–81. New York: Springer-Verlag.

INDEX

ENDNOTES

---

[1] Morin, *Seven Complex Lessons*, 21.

[2] Chaudhuri, *Integral Philosophy of Sri Aurobindo.*

[3] Chaudhuri, *Philosophy of Integralism.*

[4] Gidley, "Evolution of Consciousness as a Planetary Imperative," 224.

[5] Heehs, 165.

[6] Ibid., 177.

[7] Aurobindo, *Tales of Prison Life*, 37.

[8] Heehs, *Lives of Sri Aurobindo*, 76.

[9] Bergson, *Creative Evolution.*

[10] Ibid., 152.

[11] Heehs, *Lives of Sri Aurobindo*, 124.

[12] McIntosh, *Integral Consciousness*, 180.

[13] Gebser, *The Ever-Present Origin*, xxix.

[14] Gebser, "Cultural Philosophy as Method and Venture," 78.

[15] Feuerstein, *Structures of Consciousness*, 195.

[16] Ibid., 194–95.

[17] Ibid., 195.

[18] Ibid.

[19] Wilber, *Brief History of Everything*, 17–18.

[20] Wilber, *Integral Spirituality*, 72.

[21] Ibid., 87.

[22] Wilber, *Quantum Questions.*

[23] Wilber, *Integral Spirituality.*

[24] Kafatos, Tanzi, and Chopra, "How Consciousness Becomes the Physical Universe," 1126–27.

[25] Chalmers, "Facing Up to the Problem of Consciousness," 28.

[26] Ibid.

[27] Bohm, "Meaning and Information," 45.

[28] Ibid., 43.

[29] Bohm and Peat, *Science, Order, and Creativity*, 80.

[30] Chalmers, *Character of Consciousness*, 5.

[31] Bisson, "They're Made Out of Meat," 42.

[32] Chalmers, *Character of Consciousness,* 25.

[33] Chalmers, "Facing Up to the Problem of Consciousness," 26.

[34] Ibid., 27–28.

[35] Ibid., 27.

[36] Wheeler, *Geons, Black Holes, and Quantum Foam*, 340.

[37] Ibid.

[38] See calculation details in "Isospheres," in Chapter 5.

[39] Pockett, *The Nature of Consciousness*, xvxvi.

[40] Pockett, "Difficulties with the Electromagnetic Field Theory of Consciousness," 271.

[41] Pockett, *Nature of Consciousness*, 131–32.

[42] Ibid., 142.

[43] McFadden, "The Conscious Electromagnetic Information (CEMI) Field Theory."

[44] McFadden, "Conscious Electromagnetic Field Theory," 185.

[45] McFadden, "Synchronous Firing and Its Influence," 23.

[46] Pribram, *Brain and Perception*, 238.

[47] Pribram, "Brain and Mathematics," 219.

[48] Pribram, *Brain and Perception*, 19.

[49] Ibid., xvii.

[50] Broughton and Bryan, *Discrete Fourier Analysis and Wavelets*, 72.

[51] Pribram, *Languages of the Brain*, 26.

[52] Penfield, *Mystery of the Mind.*

[53] Ibid., 27.

[54] Ibid., 32.

[55] Neuron graphic by Jonathan Haas (2012). (**Note**: All figures not attributed to another source are proprietary to the author of this book, Shelli R. Joye). Reprinted under the terms of a Creative Commons Attribution ShareAlike 3.0 Unported license. Image retrieved from Wikimedia Commons. *Above right*, a Purkinje cell. Graphic by Santiago Ramon y Cajal (2006). Public domain image retrieved from Wikimedia Commons.

[56] Romanes, *Cunningham's Textbook of Anatomy*, 556.

[57] Netter, *Anatomy of the Nervous System*, 39.

[58] Ibid.

[59] Herculano-Houzel, "The Human Brain in Numbers," 1.

[60] Graphic by Henry Gray (1918). To the left, the groups of cells; to the right, the systems of fibers. Quite to the left of the figure a sensory nerve fiber is shown. Cell body layers are labeled on the left, and fiber layers are labeled on the right. Public domain image retrieved from Wikimedia Commons.

[61] Fields, *The Other Brain*, 72.

[62] Squire, *The History of Neuroscience*, 314.

[63] Talbot, *The Holographic Universe*, 13.

[64] Lashley as quoted in Pribram; Pribram, *The Form Within*, 22.

[65] Pribram, *The Form Within*, 24.

[66] Ibid.

[67] Köhler, *Task of Gestalt Psychology*, 10.

[68] Pribram, *The Form Within*, 35.

[69] Ibid.

[70] Ibid., 356.

[71] Ibid., 36.

[72] Netter, *Anatomy of the Nervous System.*

[73] Köhler and Henle. *Selected Papers of Wolfgang Köhler.*

[74] Pribram, *The Form Within*, 37.

[75] Ibid., 50.

[76] Ibid., 54.

[77] Ibid., 458.

[78] Ibid.

[79] Pribram, *Brain and Perception*, xvi.

[80] Pribram, *The Form Within*, 16.

[81] Pribram, "Neuropsychology of Sigmund Freud," 445.

[82] Johnston, *Holographic Visions: A History of New Science.*

[83] Gabor, "Theory of Communication."

[84] Wasson and Brieger, *Nobel Prize Winners*, 358.

[85] Johnston, *Holographic Visions*, 3.

[86] Ibid.

[87] Leith and Upatnicks, "Reconstructed Wavefronts and Communication Theory."

[88] Johnston, *Holographic Visions*, 4.

[89] Ibid., 7.

[90]Ibid.

[91] Leith and Upatnieks, "Photography by Laser."

[92] Pribram, *The Form Within*, 42.

[93] Author's note: A hologram may not necessarily be limited to visual information. Each holographic imprinting might be a complete snapshot of all current physiological, sensory, and cognitive activity at a particular moment of perception. These "snapshot holograms" of multidimensional information might be stored and chained sequentially, perhaps in layers, or filaments, or as overlapping shells as in the rings of a tree.

Such holographic memory engrams would then be available for future retrieval and playback, triggered perhaps by neuronal activity, in the selective recall of strings of specific memory sequences

[94] Lundqvist, *Nobel Lectures, Physics 1971–1980*, 82.

[95] Pribram, *Languages of the Brain*.

[96] Pribram, *The Form Within*, 140–41.

[97] Pribram, *Languages of the Brain*, 19.

[98] Ibid., 19–23.

[99] Borsellino and Poggio, "Holographic Aspects of Temporal Memory."

[100] Pribram, *Brain and Perception*, 28–29.

[101] Pribram, *The Form Within*, 84.

[102] Ibid., 69.

[103] Pribram, *Brain and Perception*, xvii.

[104] Ibid., 60.

[105] Squire, *History of Neuroscience*, 322.

[106] Pribram, *Brain and Perception*, 70.

[107] Pribram, "What Is Mind that the Brain May Order It?," 323.

[108] Pribram, *The Form Within*, 103.

[109] Gabor, "Theory of Communication."

[110] Pribram, *The Form Within*, 104.

[111] Crease, *The Great Equations*, 263.

[112] Darling, *The Universal Book of Mathematics*, 151.

[113] Alabiso and Weiss, *A Primer on Hilbert Space Theory*.

[114] Pribram, "Brain and Mathematics," 221.

[115] Gabor, "Theory of Communication," 415.

[116] Pribram, *The Form Within*, 364–65.

[117] Ibid., 516.

[118] Ibid., 338.

[119] Ibid., 142.

[120] Pribram, "What the Fuss Is All About," 29.

[121] Pribram, "Prolegomenon for a Holonomic Brain Theory."

[122] Pribram, *The Form Within*, 495.

[123] Pribram, "Brain and Mathematics," 232.

[124] Bohm, *Wholeness and the Implicate Order*.

[125] Pribram, *The Form Within,* 496.

[126] Ibid., 87.

[127] Pribram, "Brain and Mathematics," 230

[128] Pribram, "Consciousness Reassessed," 8.

[129] Ibid., 13.

[130] Ibid.

[131] Pribram, "What Is Mind that the Brain May Order It?," 322.

[132] Dewey, *The Theory of Laminated Spacetime*.

[133] Ibid., 95.

[134] Pribram, *The Form Within*, 106–7.

[135] Cazenave, *Science and Consciousness: Two Views of the Universe*.

[136] Oates, *Celebrating the Dawn: Maharishi Mahesh Yogi and the TM Technique.*

[137] Bohm and Weber, "Nature as Creativity," 35–36.

[138] Pribram, Sharafat, and Beekman, "Frequency Encoding in Motor Systems."

[139] As quoted in Squire, *History of Neuroscience*, 337.

[140] Pribram, "What Is Mind that the Brain May Order It?," 320.

[141] Ibid.

[142] Pribram, "Brain and Mathematics."

[143] Ibid., 219.

[144] Ibid., 231.

[145] Joye, *The Pribram–Bohm Holoflux Theory of Consciousness.*

[146] Romanes, *Cunningham's Textbook of Anatomy*, 840.

[147] Pribram, *The Form Within*, 109.

[148] Baggott, *The Quantum Story*, 64.

[149] Ibid., 71.

[150] As quoted in ibid., 74.

[151] Born, *The Born–Einstein Letters*, 91.

[152] Bohm, *Wholeness and the Implicate Order*, 76.

[153] Wiseman et al., "Experimental Proof of Nonlocal Wavefunction."

[154] Ibid., 1.

[155] Ibid., 3.

[156] Ibid., 5.

[157] Pribram, *The Form Within*, 408–10.

[158] Aur and Jog, *Neuroelectrodynamics: Understanding the Brain Language*, v–viii.

[159] See Pribram, "Karl Pribram: Bibliography."

[160] Pribram, "Mind, Brain, and Consciousness," 129.

[161] Ibid.

[162] Pribram, *The Form Within*, 535–36.

[163] *The Stanford Daily*, September 26, 1977: 8.

[164] Dawson, *Comprehending Cults*, 54

[165] Pribram, *Brain and Perception*, 272–73.

[166] Ibid., 173.

[167] Ibid., 275–330.

[168] Pribram, *The Form Within*, 533.

[169] Cazenave, *Science and Consciousness*.

[170] Ibid., 129–30.

[171] As quoted in Talbot, *The Holographic Universe*, 10.

[172] As quoted in Squire, *History of Neuroscience*, 342.

[173] Pribram, *The Form Within*, 538.

[174] Pribram, *Languages of the Brain.*

[175] Pribram, *The Form Within*, 531.

[176] Bohm, *Quantum Theory*, 1.

[177] Peat, *Infinite Potential.*

[178] Bohm, *Beyond Limits*, 8:00.

[179] Ibid., 12:55.

[180] Peat, *Infinite Potential*, 193–94.

[181] Ibid., 194–95

[182] Bohm and Hiley, *The Undivided Universe*, 1.

[183] Ibid., 31.

[184] Peat, *Infinite Potential*, 46.

[185] Ibid.

[186] Ibid., 47.

[187] Ibid., 62.

[188] Hiley and Peat, *Quantum Implications.*

[189] Peat, *Infinite Potential*, 57–58.

[190] As quoted in ibid., 58.

[191] Ibid., 66.

[192] Ibid., 65.

[193] Ibid., 64.

[194] Plasmas constitute over ninety-nine percent of the matter of the universe—most stars and interstellar gases exist in this fourth state of matter (Peat, *Infinite Potential*, 65).

[195] Bohm and Peat, *Science, Order, and Creativity*, xiii.

[196] Peat, *Infinite Potential*, 88.

[197] Bohm, *The Special Theory of Relativity*, vi.

[198] Hiley & Peat, "The Development of David Bohm's Ideas," 4.

[199] Bohm, *Quantum Theory.*

[200] Peat, *Infinite Potential*, 120.

[201] Ibid., 98–99.

[202] Baggott, *The Quantum Story*, 305.

[203] Peat, *Infinite Potential*, 120.

[204] Ibid., 124.

[205] Ibid., 126.

[206] Bohm, *Quantum Theory*, 1. Italics added.

[207] Ibid., 29.

[208] Peat, *Infinite Potential*, 86.

[209] Bohm, "A Suggested Interpretation of the Quantum Theory,'" 166.

[210] Garay, "Quantum Gravity and Minimum Length," 145.

[211] Peat, *Infinite Potential*, 48.

[212] Ibid., 48–49.

[213] Bohm, "Interpretation of the Quantum Theory in Terms of 'Hidden Variables.'"

[214] Peat, *Infinite Potential*, 51.

[215] Ibid., 52.

[216] Born, *The Born–Einstein Letters*, 91.

[217] Peat, *Infinite Potential*, 108.

[218] Ibid., 148.

[219] Ibid., 133.

[220] Ibid.

[221] Ibid., 168.

[222] J. T. Bell, *The Speakable and Unspeakable in Quantum Mechanics*, 128.

[223] Peat, *Infinite Potential*, 186.

[224] Bohm, *Wholeness and the Implicate Order*, 207.

[225] Ouspensky, *In Search of the Miraculous*.

[226] Peat, *Infinite Potential*, 213.

[227] Bohm and Krishnamurti, *The Limits of Thought,* vii.

[228] Peat, *Infinite Potential*, 213.

[229] See, for example, Bohm and Krishnamurti, *The Ending of Time*; and Bohm and Krishnamurti, *The Limits of Thought*.

[230] Peat, *Infinite Potential*, 227–28.

[231] Bohm and Krishnamurti, *The Limits of Thought,* vii.

[232] Peat, *Infinite Potential*, 229.

[233] As quoted in idib., 256.

[234] Ibid.

[235] Crease, *The Great Equations*, 220.

[236] Bohm, *Wholeness and the Implicate Order*, 149.

[237] Pylkkänen, *Mind, Matter, and the Implicate Order*, 21.

[238] Bohm, *Beyond Limits*, 8:00.

[239] This dimensional limit is the Planck length of 10^{-33} cm.

[240] Bohm and Hiley, *The Undivided Universe*, 9.

[241] Bohm, *Quantum Theory*, 622

[242] Ibid.

[243] Bohm and Hiley, *The Undivided Universe*, 9–10.

[244] See also Chapter 4, "The Planck Constant"; the Planck length of 10^{-33} cm was first discussed by the German physicist Max Planck, the 1918 Nobel Prize winning originator of quantum mechanics, and is derived from three values: (1) the speed of light in a vacuum, (2) the gravitational constant, and (3) the Planck constant, *h*, which is required to calculate quantum changes of wavelength as a function of temperature (Bruskiewich, *Max Planck and Black-Body Radiation*).

[245] Bohm, *Wholeness and the Implicate Order*, 190–91.

[246] Bohm, *Quantum Theory*, 18.

[247] Bohm and Peat, *Science, Order, and Creativity*, 311–12.

[248] Data from Benenson, Harris, Stocker, and Holger, *Handbook of Physics*.

[249] Ibid., 86.

[250] Gott, Jurić, Schlegel, and Hoyle, "A Map of the Universe."

[251] Bohm and Hiley, *The Undivided Universe*, 357.

[252] Yau and Nadis, *The Shape of Inner Space*, 10.

[253] Ibid., 10–11.

[254] Ibid., 12.

[255] Ibid., 13.

[256] Bohm, *Wholeness and the Implicate Order*, 190.

[257] Ibid., 186.

[258] Weber, "The Physicist and the Mystic," 187. Italics added.

[259] Hiley and Peat, eds. *Quantum Implications: Essays in Honour of David Bohm*.

[260] Ibid.

[261] Bohm, "Beyond Limits." The source is an interview with Bill Angelos; the quoted dialogue begins 32:23 into the interview. The interview runs 1:08:13 total.

[262]Globus, *The Transparent Becoming of World*, 50.

[263] Bohm, "A New Theory of the Relationship of Mind and Matter," 51.

[264] Bohm, "Meaning and Information," 45.

[265] Bohm and Peat, *Science, Order, and Creativity*, 84.

[266] Ibid.

[267] Bohm, "Meaning and Information," 45.

[268] Ibid., 2.

[269] Ibid.

[270] Bohm, "Beyond Limits," 5:30.

[271] Ibid., 1.

[272] Globus, *The Transparent Becoming of World*, 55.

[273] Peat, *Infinite Potential*, 220.

[274] Bohm, "Hidden Variables and the Implicate Order," 43.

[275] Bohm and Peat, *Science, Order, and Creativity*, 81.

[276] Ibid.

[277] Ibid.

[278] Ibid., 82. Italics added.

[279] Ibid.

[280] Bohm and Hiley, *The Undivided Universe*, 38–39.

[281] Ibid., 399.

[282] Bohm, "Meaning and Information," 45.

[283] Ibid., 46.

[284] Bohm, *Wholeness and the Implicate Order*, 196–97.

[285] Ibid., 200.

[286] Weber, "The Physicist and the Mystic," 196.

[287] Bohm, "The Implicate Order and the Super-Implicate Order," 45–46.

[288] Bohm, *The Undivided Universe*, 389–90. Italics added.

[289] Kant, *Critique of Pure Reason*, 335.

[290] Graphic by Dewynne (2012). Reprinted under the terms of a Creative Commons CC0 1.0 Universal Public Domain Dedication. Image retrieved from Wikimedia Commons.

[291] Bohm, *Beyond Limits*, 5:30.

[292] Peat, *Infinite Potential: The Life and Times of David Bohm*.

[293] Bohm, "The Enfolding-Unfolding Universe," 62.

[294] Bohm and Hiley, *The Undivided Universe*, 381.

[295] Ibid., 381–82.

[296] Bohm, "Beyond Limits," 7:20.

[297] Carr and Giddings, "Quantum Black Holes," 32.

[298] Wong, *The Shambhala Guide to Taoism*, 124–31.

[299] Globus, *Quantum Closures and Disclosures*, 175.

[300] *Left image*: Graphic by Bain (2012). Public domain image under the terms of Title 17, Chapter 1, Section 105 of the US Code; *Right image*: Graphic by Fulvio314 (2013). Reprinted under the terms of a Creative Commons Attribution ShareAlike 3.0 Unported license. Image retrieved from Wikimedia Commons.

[301] Eliot, *Collected Poems 1909—1962*, 176.

[302] Talbot, *The Holographic Universe*, 54.

[303] Neville, "Saral and David Bohm," picture 8.

[304] Bohm, *Wholeness and the Implicate Order*, 193.

[305]Graphic by Jbarta (2014). This work is in the public domain under the terms of Title 17, Chapter 1, Section 105 of the US Code.

[306] Baggott, *The Quantum Story*, 7–8.

[307] Ibid., 15.

[308] Ibid., 15.

[309] Planck, "On the Law of Distribution of Energy in the Normal Spectrum."

[310] Ibid.

[311] Baggott, *The Quantum Story*, 11.

[312] Benenson, Harris, Stocker, and Holger, *Handbook of Physics*.

[313] Baggott, *The Quantum Story*, 21.

[314] Wheeler, *A Journey into Gravity and Spacetime*.

[315] Zizzi, "Consciousness and Logic," 459–60.

[316] Ledrew, "The Real Starry Sky."

[317] Benenson, Harris, Stocker, and Holger, *Handbook of Physics*.

[318] Bohm, *Wholeness and the Implicate Order*, 190.

[319] Bohm, "Hidden Variables and the Implicate Order," 44.

[320] Ibid., 100.

[321] Zizzi, "Consciousness and Logic," 469.

[322] Ibid., 468.

[323]Dimensional data from Benenson, Harris, Stocker, and Holger, *Handbook of Physics*.

[324] Eliot, *Four Quartets*, 16.

[325] Zizzi, "Consciousness and Logic," 468–69.

[326] Wheeler, *A Journey into Gravity and Spacetime*, 222.

[327] Van Dokkum and Conroy, "A Substantial Population of Low-Mass Stars."

[328] Ibid.

[329] Bohm, *Wholeness and the Implicate Order*, 191–92.

[330] Ibid., 191.

[331] Kuehn, *A Student's Guide Through the Great Physics Texts*, 332.

[332] Gebser, *The Ever-Present Origin: Part One: Foundations of the Aperspectival World*.

[333] Eliot, *Four Quartets*, 15.

[334] Eliot, *Collected Poems 1909–1962*, 126.

[335] Teilhard, "Life and the Planets," 122. Essay written in 1945.

[336] Bohm, *Wholeness and the Implicate Order*, 192.

[337] Edmondson, *Synergetic Geometry of Buckminster Fuller*, 19.

[338] Rendek, Fromme, Grotjohann, and Fromme, "Self-Assembled Three-Dimensional DNA Nanostructure," 141.

[339] Bohm, *Quantum Theory*, 6.

[340] Wheeler, *A Journey into Gravity and Spacetime*, v.

[341] Baggott, *The Quantum Story*, 242.

[342] Wheeler, *Geons, Black Holes, and Quantum Foam*, 338.

[343] Modified from figure captioned, "Does looking back 'now' give reality to what happened 'then'?"; see Wheeler, *Geons, Black Holes, and Quantum Foam*, 339.

[344] Bohm, *A Question of Physics: Conversations in Physics and Biology*, 132.

[345] Pribram, *The Form Within.*

[346] Globus, *The Transparent Becoming of World*, 132.

[347] Wheeler, *A Journey into Gravity and Spacetime*, 223.

[348] Bohm, *Wholeness and the Implicate Order*, 207.

[349] Eliot, *Collected Poems 1909–1962,* 126.

[350] Dewey, *Theory of Laminated Spacetime.*

[351] Bohm and Hiley, *The Undivided Universe*, 9–10.

[352] Susskind, *The Black Hole War*, 173.

[353] Ibid.

[354] Wheeler, *Information, Physics, Quantum*, 361.

[355] Wheeler, *Geons, Black Holes & Quantum Foam*, 297.

[356] Wheeler, *A Journey into Gravity and Spacetime*, 222.

[357] Ibid.

[358] Bekenstein, "Black Holes and Entropy."

[359]Wheeler, *A Journey into Gravity and Spacetime*, 220.

[360] Zizzi, "Consciousness and Logic," 460.

[361] Crease, *The Great Equations*, 135.

[362] Cook, *The Theory of the Electromagnetic Field.*

[363]Graphic by Tjlaxs (2005). Public domain image published in the US before 1923. Retrieved from Wikimedia Commons.

[364] Hertz, *Electric Waves.*

[365] E. T. Bell, *Men of Mathematics*, 16.

[366] Ruppert, *History of the International Electrotechnical*

Commission.

[367] Pribram, *The Form Within*, 105.

[368] Wilber, *The Holographic Paradigm*, 181.

[369] Merrell-Wolff, *The Philosophy of Consciousness Without an Object.*

[370] Kuo, *Network Analysis and Synthesis*, 13.

[371] Ibid.

[372] Ibid., 3.

[373] Bohm, *Quantum Theory*, 1.

[374] Dorf, *The Electrical Engineering Handbook*, 1538.

[375] Planck time is a constant set by the speed of light and the gravitational constant, at a value of 5.39106×10^{-44} seconds; Runehov and Oviedo, eds., *Encyclopedia of Sciences and Religions*, 1772.

[376] Hameroff and Penrose, "Conscious Events as Orchestrated Space–Time Selections."

[377] Hameroff, Craddock, and Tuszynski, "Quantum Effects in the Understanding of Consciousness," 230.

[378] Jibu and Yasue, *Quantum Brain Dynamics and Consciousness*, 147.

[379] Stuart, Takahashi, and Umezawa, "Mixed System Brain Dynamics."

[380] Jibu and Yasue, "Quantum Brain Dynamics and Quantum Field Theory," 288.

[381] Chen, *Introduction to Plasma Physics*, 56.

[382] Jibu and Yasue, *Quantum Brain Dynamics and Consciousness*, 164–65.

[383] Jibu and Yasue, "Quantum Brain Dynamics and Quantum Field Theory," 293.

[384] Chen, *Introduction to Plasma Physics*, 1.

[385] Jibu and Yasue, "Quantum Brain Dynamics and Quantum Field Theory," 293.

[386] Dorf, *The Electrical Engineering Handbook*, 27.

[387] McCraty, *The Energetic Heart.*

[388]Graphic by Svdmolen (2005). This work is in the public domain under the terms of Title 17, Chapter 1, Section 105 of the US Code. Image retrieved from Wikimedia Commons.

[389] Turgeon, *Clinical Hematology.*

[390] Romanes, *Cunningham's Textbook of Anatomy.*

[391] Wick, Pinggera, and Lehmann, *Clinical Aspects and Laboratory Iron Metabolism*, 6.

[392] Nishikawa and Wakatani, *Plasma Physics.*

[393] Wheeler, *A Journey into Gravity and Spacetime*, 222.

[394] Romanes, *Cunningham's Textbook of Anatomy*, 137.

[395] Wheeler, *A Journey into Gravity and Spacetime*, 222.

[396] McCraty, *The Energetic Heart*, 1.

[397] The carbon dioxide laser is one of the earliest gas lasers to be developed (invented by Kumar Patel of Bell Labs in 1964), and is still one of the most useful. Carbon dioxide lasers are the highest-power continuous wave lasers that are currently available; see Milonni and Eberly, *Laser Physics.*

[398] Clark, *Photography by Infrared*, 78.

[399] Pizzi et al., "Evidences of New Biophysical Properties of Microtubules."

[400] Graphic by Glatzmaier (2007). This work is in the public domain under the terms of Title 17, Chapter 1, Section 105 of the US Code. Image retrieved from Wikimedia Commons.

[401] Lilly, *The Deep Self: Consciousness Exploration in the Isolation Tank.*

[402] Booth, Koren, and Persinger, "Increased Feelings of the Sensed Presence During Exposures to Weak Magnetic Fields."

[403] Allaby and Allaby, "Telluric Current."

[404] Ibid.

[405] Cheney, *Tesla: Man Out of Time.*

[406] Ibid, 78.

[407] Vejvoda, *17th International Conference on Reactor Technology*, 17.

[408] Ruhenstroth-Bauer, "Influence of the Earth's Magnetic Field," 336.

[409] Wiltschko and Wiltschko, "Magnetoreception in Birds and Other Animals," 675.

[410] Ibid, 677.

[411] Ruhenstroth-Bauer, "Influence of Earth's Magnetic Field," 195.

[412] Persinger, "Schumann Resonance Frequencies," 25.

[413] Sentman, "Schumann Resonances," 271.

[414] Allaby and Allaby, "Telluric Current," 43.

[415] Persinger, "Schumann Resonance Frequencies," 25.

[416] Warren and Brandt, "Optical Constants of Ice from the Ultraviolet to the Microwave."

[417] Dorf, *The Electrical Engineering Handbook.*

[418] Graphic by Gray (1918). Reprinted under the terms of a Creative Commons Attribution ShareAlike 3.0 Unported license. Image retrieved from Wikimedia Commons.

[419] National Multiple Sclerosis Society. "Cerebrospinal Fluid (CSF)."

[420] Graphic by Gray (1918). Reprinted under the terms of a Creative Commons Attribution ShareAlike 3.0 Unported license. Image retrieved from Wikimedia Commons.

[421] Graphic by Averse (2007).

[422] Collister, *A Journey in Search of Wholeness and Meaning*, 69.

[423] Bohm, *Wholeness and the Implicate Order.*

[424] Chalmers, "Facing Up to the Problem of Consciousness."

[425] Block, *Consciousness, Function, and Representation.*

[426] Bohm, *Wholeness and the Implicate Order.*

[427] Teilhard de Chardin, "The Nature of the Point Omega."

[428] László , *The Self-Actualizing Cosmos.*

[429] Pribram, *Brain and Perception*, 70.

[430] Wheeler, *Geons, Black Holes, and Quantum Foam*, 339; also see Fig. 21.

[431] Nilson and Hornung, *Practical Radio Communication,* 54.

[432] Zimmer, "Origins: On the Origin of Eukaryotes," 667.

[433] Dennett, *Consciousness Explained*, 406.

[434] Oyster, *The Human Eye: Structure and Function.*

[435] Data from Geiger, Aron, and Todhunter, *The Climate Near the Ground*, 452.

[436] Calculations in this paragraph use Wien's Law to convert wavelength to frequency.

[437] Cohn, *Near-Infrared Spectroscopy: Potential Clinical Benefits in Surgery*, 323.

[438] Hameroff, "Is Your Brain Really a Computer," pars. 5, 6, 8, 9.

[439] Pizzi et al., "Evidence of New Biophysical Properties of Microtubules," 1.

[440] Ibid., 13.

[441] Havelka et al., "High-Frequency Electric Field and Radiation Characteristics of Cellular Microtubule Network," 31.

[442] Ishikawa and Marshall, "Ciliogenesis: Building the Cell's Antenna," 222.

[443] Annotations added by author to graphic by Jahoe (2012). Reprinted under the terms of a Creative Commons Attribution ShareAlike 2.5 Generic license. Image retrieved from

Wikimedia Commons.

[444] Drachman, "Do We Have Brain to Spare?", 1.

[445] Lilly, *The Mind of the Dolphin: A Nonhuman Intelligence*.

[446] Teilhard de Chardin, *The Human Phenomenon*.

Made in United States
Orlando, FL
11 November 2026

90609407R00115